AQA Level 2 Certificate in Furt

By Vali Nasser

Copyright © 2016

E-book editions are also available for this title. For more information email:

valinasser@gmail.com

All rights reserved by the author. No part of this publication can be reproduced, stored in a retrieval system, or transmitted in any form or by any means, electronic, mechanical, photocopying, recording or otherwise, without the prior permission of the publisher and/or author.

February 2016

ISBN-13: 978-1530207510

ISBN-10: 1530207517

Every effort has been made by the author to ensure that the material in this book is up to date and in line with the requirements to pass the AQA Level 2 certificate in Further Mathematic at the time of publication. The author will also do his best to review, revise and update this material periodically as necessary. However, neither the author nor the publisher can accept responsibility for loss or damage resulting from the material in this book

INTRODUCTION	8
NUMBER SECTION 1	**10**
Fractions, decimals and percentages	10
Simple Interest and Compound Interest	15
FRACTIONS	**18**
Simplifying fractions	18
Finding fraction of an amount	19
Adding and Subtracting Fractions	20
Adding and subtracting mixed numbers	23
Multiplying Fractions	23
Division of Fractions	24
Multiplying mixed numbers together	25
Dividing mixed numbers together	26
PROPORTIONS AND RATIOS	**27**
Scales and ratios	29
NUMBER WORK SECTION 1: PRACTICE QUESTIONS	**30**
ANSWERS TO NUMBER WORK SECTION 1: PRACTICE QUESTIONS	**32**

NUMBER SECTION 2 33

Indices and Surds 33

Fractional Indices 34

Rational and Irrational Numbers 34

Surds 36

NUMBER WORK SECTION 2: PRACTICE QUESTIONS 37

ANSWERS TO NUMBER WORK SECTION 2 38

ALGEBRA SECTION 1 39

Algebra basics 39

Reminder: Multiplying positive and negative numbers. 39

Dividing positive and negative numbers. 39

Simplifying algebraic expressions 40

Multiplying out brackets 41

SIMPLIFYING ALGEBRAIC FRACTIONS 45

ALGEBRA PRACTICE QUESTIONS - SIMPLIFYING EXPRESSIONS
.... 46

ANSWERS TO SIMPLIFYING EXPRESSIONS 47

FACTORISING 48

PRACTICE QUESTIONS ON FACTORISING ALGEBRAIC EXPRESSIONS: 49

ANSWERS TO FACTORISING EXPRESSIONS: 50

ALGEBRAIC SUBSTITUTION AND FORMULA 51

ALGEBRAIC PROOFS 53

ANSWERS TO ALGEBRAIC PROOFS: 55

CHANGING THE SUBJECT 57

PRACTICE QUESTIONS ON CHANGE THE SUBJECT 59

ANSWERS TO CHANGE THE SUBJECT: 60

SEQUENCES 61

QUADRATIC SEQUENCE 63

ANSWERS TO QUESTIONS ON SEQUENCES: 65

ALGEBRA SECTION 2 66

Solving equations 66

SOLVING WORD PROBLEMS USING ALGEBRA 69

SIMULTANEOUS EQUATIONS 71

PRACTICE QUESTIONS ALGEBRA SECTION 2 75

ANSWERS TO ALGEBRA SECTION 2 77

ALGEBRA SECTION 3 78

Solving Quadratic Equations 78

SOLVING QUADRATIC INEQUALITIES 83

GRAPHS OF QUADRATIC EQUATIONS 85

CUBIC EQUATION 88

EXPONENTIAL GRAPHS 89

SOLVING EQUATIONS USING GRAPHICAL METHODS 90

Solving equations mathematically when one is linear and the other is quadratic: 90

REMAINDER AND FACTOR THEOREM 93

Remainder Theorem 93

Factor Theorem 93

SOLVING CUBIC EQUATIONS 95

PRACTICE QUESTIONS ALGEBRA SECTION 3 98

ANSWERS TO ALGEBRA SECTION 3 99

GEOMETRY 102

PYTHAGORAS' THEOREM 108

CIRCLE THEOREM 110

EQUATION OF A CIRCLE 114

PRACTICE QUESTIONS ON EQUATION OF A CIRCLE 116

ANSWERS TO EQUATION OF CIRCLE QUESTIONS.................117

AREAS AND VOLUMES OF COMMON SHAPES 119

Perimeters, Areas and Volumes of common shapes 119

LINEAR EQUATIONS..126

USING RATIOS TO FIND CO-ORDINATES 129

WORKING OUT EQUATIONS OF 'NORMALS' AND 'PARALLEL' LINES..130

Finding parallel lines..131

PRACTICE QUESTIONS IN GEOMETRY AND CO-ORDINATE GEOMETRY..132

ANSWERS..134

TRIG FOR RIGHT ANGLED TRIANGLES..........................137

TRIG FOR NON- RIGHT ANGLED TRIANGLES 140

GRAPHS OF TRIG FUNCTIONS 142

TRIG IDENTITIES 145

CALCULUS..146

Differentiation..146

ANSWERS TO PRACTICE QUESTIONS IN TRIGONOMETRY AND CALCULUS..151

A MATRIX IS SIMPLY AN ARRAY OF NUMBERS **156**

MULTIPLYING A MATRIX BY ANOTHER MATRIX **158**

PRACTICE QUESTIONS ON MATRICES **163**

ANSWERS TO PRACTICE QUESTIONS ON MATRICES **164**

Introduction

This book 'AQA Level 2 certificate in Further Mathematics' has many examples and exam type practice questions that will help you get up to speed with a higher level of mathematics than GCSE. This qualification fills the gap for high achieving students by assessing their higher order mathematical skills, particularly in algebraic reasoning which will stretch and challenge you as well as help you to build on the topics in Key Stage 4. **It is an additional qualification to the GCSE Mathematics, rather than a replacement.** This new qualification places an emphasis on higher order proficiency, rigorous argument and problem solving skills. It also gives an introduction to calculus and matrices and develops further skills in trigonometry, functions and graphs. The AQA Level 2 Certificate in Further Mathematics is for learners who either already have or are expected to achieve grades A and A* in GCSE mathematics and are likely to progress to study A-Level mathematics and possibly further mathematics.

For your information AQA Level 2 certificate in Further Mathematics has two papers in the exam. Paper 1 (non-calculator) which lasts 1 hour 30 minutes and consists of 70 marks. Paper 2 is calculator based, lasts 2 hours and has 105 marks allocated.

This qualification is designed to be taught: either in parallel with GCSE Mathematics, after GCSE Mathematics or instead of GCSE Mathematics if the assumed knowledge areas of this specification have been covered.

About the Author

The author of this book has experience in both consultancy work and teaching. As a specialist mathematics teacher he has tutored and taught mathematics in schools as well as in adult education. The author's initial book 'Speed Mathematics Using the Vedic System' has a significant following and has been translated into Japanese and Chinese as well as German. His book 'Pass the QTS Numeracy Skills Test with Ease' has been very popular with teacher trainees and 'Pass the Numerical Reasoning Tests' is popular with graduates wanting to further their career. He is also the author of several GCSE books. He hopes that his new book **AQA Level 2 Certificate in Further Mathematics** will be helpful to pupils who need to revise the topics for this particular exam.

Number Section 1

Fractions, decimals and percentages

I am sure most of you are aware that $\frac{1}{2}=0.5$. This in turn is equal to 50%.

It is worth reviewing this fact. In addition, you should try and remember the following other equivalences if you have forgotten them:

Fractions, decimals and percentage equivalents

Fractions	Decimal	Percentage
$\frac{1}{2}$	0.5	50%
$\frac{1}{4}$	0.25	25%
$\frac{3}{4}$	0.75	75%
$\frac{1}{10}$	0.1	10%
$\frac{1}{5}$	0.2	20%

If, we know $\frac{1}{2} = 0.5$

We can deduce that $\frac{1}{4} = \mathbf{0.25}$

(Since a quarter is half of half)

Similarly $\frac{1}{8}$ is **0.125**

We can do this quickly because all we do is halve each decimal value.

Half of 0.5 is 0.25, Half of 0.25 is 0.125

We can of course continue this process.

Further if we know $\frac{1}{10} = 0.1$ **we can now work out** $\frac{2}{10}, \frac{3}{10}, \frac{7}{10}$ **etc.**

$\frac{2}{10} = 0.2$ (2 × 0.1), $\frac{3}{10} = 0.3$ (3 × 0.1, $\frac{7}{10} = 0.7$ (7 × 0.1), $\frac{9}{10} = 0.9$ (9 × 0.1)

Another useful fraction and decimal equivalent to remember is $\frac{1}{3}$ =0.333… (0.3 recurring)

The key equivalent percentages to remember are as follows:

$\frac{3}{4} = 75\%, \frac{1}{2} = 50\%, \frac{1}{4} = 25\%, \frac{1}{8} = 12.5\%, \frac{1}{10} = 10\%$

See summary box below

Summary:

Remember the following equivalences

$\frac{1}{2} = 0.5$ = 50%, $\frac{1}{4} = 0.25 = 25\%$, $\frac{3}{4} = 0.75 = 75\%$, $\frac{1}{10} = 0.1 = 10\%$

Also if you can try to remember, $\frac{1}{5} = 0.2 = 20\%$, and $\frac{2}{5} = 0.4 = 40\%$, $\frac{1}{3} = 0.333\ldots$ (0.3 recurring) = 33.33% (to 2 decimal places)

To convert a fraction into a percentage, simply multiply the fraction by **100**

Working out increase or decrease in percentages from original value

Example1: In a certain corner shop 16 packs of cereal A were sold in week 1. In the same shop 20 packs of the same cereal were sold in week 2. What was the percentage increase in the cereal packs A sold from week 1 to week 2?

Method: Increase in number of cereal packs A = 20 – 16 =4. Original number of cereal packs =16. The increase of 4 was based on 16 cereal packs. To work out the percentage increase we simply divide the increase by the original number of cereal packs and multiply this by 100. That is $\frac{4}{16} \times 100 = \frac{1}{4} \times 100 = 25\%$

To work out decrease in percentages (uses the same principle as above)

Example2: The original price of a projector was: £150, the new price is reduced to £135. What is the percentage decrease in price? The decrease in price is £150 - £135 = £15. The decrease over the original price is $\frac{15}{150}$. To turn this into a percentage we multiply $\frac{15}{150} \times 100 = \frac{1500}{150} = 10\%$. So the decrease in percentage price is 10%.

The basic formula to work out increase or decrease percentage change is shown below:

$$\frac{difference\ between\ final\ and\ original\ value}{original\ value} \times 100$$

One thing to remember though is that the increase or decrease in **percentage points** is different from **increase or decrease in percentages**.

To illustrate this consider the example below:

The unemployment rate in a region A was 8% in 2010. In 2011 the unemployment rate in the same region was 10%. **(1)** What was the **percentage point** increase in unemployment from 2010 to 2011? **(2)** What is the **percentage increase** in unemployment from 2010 to 2011?

(1) The **percentage point** increase is simply 2% (i.e. from 8% to 10%)

(2) However the **percentage increase** in unemployment is $\frac{2}{8} \times 100 = \frac{200}{8} = \frac{100}{4} = 25\%$.

In the Numerical reasoning test context, if you are asked to work out the **percentage point increase,** say in sales in Product A changing from 20% to 30%. The answer is obviously 10%. But if asked to work out the **percentage increase**, then the answer is $\frac{10}{20} \times 00 = \frac{1000}{20} = \frac{100}{2} = 50\%$

Miscellaneous questions involving fractions and percentages

Example 1: Finding a fraction and turning it into a percentage

There are 40 builders in a small town in Yorkshire. In a particular month 5 builders are without work. What is the percentage of builders that do not have work in this month in this small town?

The fraction of builders without work = $\frac{5}{40}$, by dividing top and bottom numbers by 5 we get $\frac{1}{8}$.

To convert 1/8 into a percentage simply multiply 1/8 by 100

= 1/8 X100 =100/8 = 50/4 =25/2 =12.5%.

(Another method: We know ¼ =25%. Hence 1/8 =12.5% (Since 1/8 is half of a quarter)

Simple Interest and Compound Interest

Simple Interest

Example 1: I have £5000 in a building society account which pays me simple interest of 3% per annum. I keep my money for 3 years. How much in total will I have at the end of the 3 year period?

Method: At the end of the first year the total interest I will receive is 3% of £5000. This is $3\frac{3}{100} \times 5000 = \frac{15000}{100} = £150$ per annum.

At the end of 3 years I will receive 3X £150 = £450 total interest. This means the total amount I will have is £5450 (original £5000 plus 3 years of simple interest)

You can if you like use the formula below to work out the total simple interest over a given period of time.

I = PRT where I = Total interest, P = Principal amount (original amount), R is the annual interest rate and T = the time in years.

So in the above case $I = 5000 \times \frac{3}{100} \times 3 = £5000 \times \frac{9}{100} = £450$

Finally, to find the total amount we have at the end of the 3 year period, we simply add £450 to the original £5000 to get £5450

Example2: Work out the final amount at the end of one year if there is a 10% increase per annum and I have $3000 to start with.

The traditional method is to work out 10% of $3000 first. Then add this answer to $3000 to get the final answer.

So 10% of $3000 = $300. So the final price after a 10% increase is $3000 + $300 = $3300

Here is fast and efficient method to work out the final price:

15

Simply work out 1.1× 000

Since 1.1 denotes a 10% increase.

Why 1.1? Since 100% plus 10% = 1 + 0.1 = 1.1

Now 1.1× 3000 = $3300 which is the final answer

Compound Interest (you can use a calculator to work out the examples below!)

Now consider a problem involving recurring percentage changes

Example 1

Find the value of $5000 if I gain a profit of 10% the first year followed by (10% of the new amount) in the second year. (This is called compound interest)

THIS MEANS THE INCREASE IS 1.1× FOLLOWED BY 1.1× AGAIN

Or $(1.1)^2 \times 5000$

$(1.1)^2 \times 5000 = 1.21 \times 5000 = \$6050$

So the final value is $6050

Example 2

I buy a one bedroom apartment for $200,000. It increases in value by 5% per annum

How much will it be worth in 15 years?

Method: Increase after 1 year will be 1.05×$200,000, after two years it will be: $(1.05)^2 \times \$200{,}000$, after three years it will be $(1.05)^3 \times \$200{,}000$. So, after 15 years it will be worth $(1.05)^{15} \times \$200{,}000 = \$415786$

Example 4:

A car depreciates by 30% per annum. I buy it at £18000. What is its value in 5 years' time? Give your answer to the nearest pound

Method:

After one year its value will decrease by 30%, so its new value will be 70% of original as shown below:

£18000×0.7, hence, after five years its value will be $18000× $(0.7)^5$

Value after five years is £3025

Example 5:

I have a house that is currently valued at £300,000. House prices are predicted to fall 7% per annum for the next two years followed by a 5% growth for a subsequent five years. What is the value of my house after seven years?

Method:

The value of my house changes by 'depreciating' for two years followed by 'appreciating' for the subsequent five years.

So the value is £300000 × $(0.93)^2$ × $(1.05)^5$

= £331157 (to the nearest pound)

(Note: 0.93 denotes the depreciation by 7%, and 1.05 denotes the growth by 5%)

Fractions

Simplifying fractions

Reducing a fraction to its lowest terms

Basically you need to find numbers that divide into the top number (numerator) as well as the bottom number (denominator), and then divide them both by the same number (start with 2, if doesn't go then choose 3, then 5, and then the next prime factor e.g. 7, 11, etc.)

Example1: Reduce $\frac{16}{24}$ to its lowest terms.

8 divides exactly into 16 and 24, so in the fraction $\frac{16}{24}$ divide top and bottom by 8. This gives the answer as $\frac{2}{3}$

In case you can't see this straight away, try starting with the number two and work your way numerically upwards using the next prime factor i.e. try 3, then 5 etc. if required

So for the fraction $\frac{16}{24}$ we can start dividing top and bottom by 2 to give us $\frac{8}{12}$, then do the same again as both 8 and 12 are still divisible by 2. This gives us $\frac{4}{6}$ and finally repeating the process once more reduces the fraction to $\frac{2}{3}$ which is the simplest form.

Example 2

Simplify $\frac{9}{12}$ to its lowest terms

In this case we can't divide top and bottom by 2, so we try 3. Since 3 will go into both 9 and 12, we can reduce this to the fraction $\frac{3}{4}$ (since 9 ÷3 =3 and 12 ÷ 3 =4)

Hence, $\frac{9}{12}$ reduces to $\frac{3}{4}$

Example 3:

Reduce fraction $\frac{49}{77}$ to its lowest terms. This time we need to spot that 2, 3, 5, does not go into either 49, or 77. Either by trial and error or by spotting the right number we notice 7 goes into both the numerator and the denominator. This reduces

$\frac{49}{77}$ to $\frac{7}{11}$

Cancelling down fractions to its simplest form (lowest terms)

To simplify a fraction to its lowest terms you divide the numerator and the denominator by the same prime factors (2, 3, 5, 7, 11, etc.) to give the equivalent fractions as shown in the examples above

---

Finding fraction of an amount

Example1: Find $\frac{2}{5}$ of 25, simply replace the 'of' by X. (times)

So $\frac{2}{5}$ of 25 becomes $\frac{2}{5}$ X 25

To work this out find out 1/5 of 25 and then multiply the answer by 2. So 25 divided by 5, equals 5, then 2 x 5 =10. Hence $\frac{2}{5}$ of 25 =10

> **Example:**
>
> 60 people apply for a certain job vacancy. 12 people are short-listed for an interview. What is the proportion of people that are not short listed for the interview? Give your answer as a decimal.
>
> Total number of people applying for this job = 60, since 12 people are shortlisted, this means 48 are not shortlisted. Hence the proportion that is not shortlisted = 48/60. If you divide top and bottom by 6, this simplifies to 8/10.
>
> The answer as a decimal is 0.8

Adding and Subtracting Fractions

This next section will help you revise adding, subtracting, multiplying and dividing fractions together

Consider adding and subtracting fractions together.

When the bottom numbers (denominators) are the same, just add the top numbers together keeping the bottom number the same, likewise for subtraction just subtract the top two numbers.

Example 1: $\dfrac{2}{5} + \dfrac{1}{5} = \dfrac{3}{5}$

Example 2: $\dfrac{2}{5} - \dfrac{1}{5} = \dfrac{1}{5}$

When the denominators are different

Example 3: Work out $\dfrac{1}{2} + \dfrac{2}{5}$

When the denominators are different, the traditional method of doing this is to find the lowest common denominator. We have to find a number that both 2 and 5 will go into. This is clearly 10.

We can now re-write the fraction with the same common denominator.

To do this we have to ask how did we get the denominator from 2 to 10 for the first part, and likewise for the second part from 5 to 10. The answer is shown below:

$$\frac{1 X 5}{2 X 5} + \frac{2 X 2}{5 X 2} = \frac{5}{10} + \frac{4}{10} = \frac{9}{10}$$

We had to multiply top and bottom by 5 for the first part and top and bottom by 2 for the second part as shown above. We can then add the fraction as we have the same common denominator.

We can however use another very simple strategy that always works. The method is that of crosswise multiplication.

The basic method is to take the fraction sum and do crosswise multiplication as shown by the arrows. In addition, multiply the denominators (bottom numbers) together to get the new denominator.

Example1: $\frac{1}{2} + \frac{2}{5} = \frac{1}{2} \times \frac{2}{5} = \frac{1X5 + 2X2}{2X5} = \frac{5+4}{10} = \frac{9}{10}$

We notice that if we cross multiply as shown we get 1 X 5 and 2 X 2 respectively at the top. To get the bottom number we simply multiply the bottom numbers, 2 and 5 together. So the denominator is 2 X 5=10.

Let us try another example:

Example2: Work out $\frac{3}{7} + \frac{2}{5}$

Using crosswise multiplication and adding rule, as well as multiplying the bottom two numbers we get:

$$\frac{3}{7} \times \frac{2}{5} = \frac{3X5 + 7X2}{35} = \frac{15+14}{35} = \frac{29}{35}$$

This is a very elegant method which always works

Example 3: Work out $\dfrac{3}{7} - \dfrac{2}{5}$

This is similar to the above except instead of adding we now subtract as shown below.

$$\dfrac{3}{7} \times \dfrac{2}{5} = \dfrac{3X5 - 7X2}{35} = \dfrac{15 - 14}{35} = \dfrac{1}{35}$$

Note: In fact you can use this method when adding or subtracting any fraction that you find difficult. Even if you use this method for simple cases, you will still get the right answer but you may have to cancel down to get the lowest terms for the final answer.

For example we know that $\dfrac{1}{4} + \dfrac{1}{2} = \dfrac{3}{4}$

But if we didn't know and used the method shown we would get

$\dfrac{1}{4} \times \dfrac{1}{2} = \dfrac{1X2 + 4X1}{4X2} = \dfrac{2+4}{8} = \dfrac{6}{8} = \dfrac{3}{4}$ (we get this by dividing both the numerator and denominator in $\dfrac{6}{8}$ by 2). So we get the same answer in the end

Question involving fractions

(1) Find $2\dfrac{3}{4}$ of £64

We first work out 2 X 64 = 128, to work out three quarters of 64 we first work out a half and then add it to a quarter of 64.

Half of £64 is £32

A quarter of £64 (is half of £32) is £16

Hence three quarters of £64 = £32 + £16 = £48

So two and three quarters of £64 = £128 + £48 = £176

Adding and subtracting mixed numbers

We first add or subtract the whole numbers and then the fractional parts.

Ex1: $2\frac{2}{5} + 4\frac{3}{7}$

Adding the whole numbers we get 6. (Simply add 2 and 4)

Now add the fractional parts to get: $\frac{14+15}{35} = \frac{29}{35}$

So the answer is $6\frac{29}{35}$

Ex2: $4\frac{3}{7} - 2\frac{2}{5}$

Subtract the whole numbers and then the fractional parts, which gives us:

$2\frac{15-14}{35} = 2\frac{1}{35}$

Multiplying Fractions

Multiplying fractions by the traditional method is quite efficient so we will consider only this approach.

Example 1: $\frac{2}{3} \times \frac{5}{7} = \frac{10}{21}$

In this case we simply multiply the top two numbers to get the new numerator and multiply the bottom two numbers together to get the new denominator, as shown above.

Another example will help consolidate this process:

Example 2: $\dfrac{10}{21} \times \dfrac{5}{7} = \dfrac{50}{147}$

(Multiply 10 × 5 to get 50 for the numerator and 21 × 7 to get 147 for the denominator)

Division of Fractions

When dividing fractions we invert the second fraction and multiply as shown.

Think of an obvious example. If we have to divide ½ by ¼ we intuitively know that the answer is 2. The reason for this is that there are 2 quarters in one half. Let us see how this works in practice.

Example 1: $\dfrac{1}{2} \div \dfrac{1}{4} = \dfrac{1}{2} \times \dfrac{4}{1} = \dfrac{4}{2} = 2$

Step 1: Re-write fraction as a multiplication sum with the second fraction inverted.

Step 2: Work out the fraction as a normal multiplication

Step 3: Simplify if possible. In this case 4 divided by 2 is 2.

Example 2: $\dfrac{6}{11} \div \dfrac{5}{11} = \dfrac{6}{11} \times \dfrac{11}{5} = \dfrac{66}{55} = \dfrac{6}{5} = 1\dfrac{1}{5}$

Step1: Re-write the fraction inverting the second fraction as shown

Step2: Multiply the top part and the bottom part to get $\dfrac{66}{55}$ as shown.

Step 3: Simplify this by dividing top and bottom by 11 to get $\frac{6}{5}$.

Now this finally simplifies to $1\frac{1}{5}$ as shown.

The following steps are required to convert a mixed number into a fraction. Consider the mixed fraction $2\frac{1}{4}$.

Step 1: Multiply the denominator of the fractional part by the whole number and add the numerator. In this case this works out to 2 × 4 + 1 = 9. This now becomes the new numerator.

Step 2: The denominator stays the same as before. Now re-write the new fraction as $\frac{9}{4}$. (That is the new numerator ÷ existing denominator)

Let us look at another example. Convert the mixed number, $3\frac{3}{7}$ into a fraction.

Step 1: Multiply denominator of fractional part by whole number and add the numerator.

This gives 3 × 7+3 = 24 as the new numerator.

Step2: Re-write fraction as new fraction. This is now the new numerator ÷existing denominator. This gives us $\frac{24}{7}$

Multiplying mixed numbers together

Consider the examples below:

Example: $1\frac{1}{5} \times 1\frac{3}{8}$

The method is simply to convert both mixed numbers into fractions and multiply as shown below:

$$1\frac{1}{5} \times 1\frac{3}{8} = \frac{6}{5} \times \frac{11}{8} = \frac{66}{40} = 1\frac{26}{40} = 1\frac{13}{20}$$

(Notice $\frac{26}{40}$ simplifies to $\frac{13}{20}$)

Dividing mixed numbers together

Example: $1\frac{1}{2} \div 1\frac{1}{4}$

There are two steps required to work out the division of mixed numbers.

Step1: Convert both mixed numbers into fractions as before

Step 2: Multiply the fractions together but invert the second one.

$$1\frac{1}{2} \div 1\frac{1}{4} = \frac{3}{2} \div \frac{5}{4} = \frac{3}{2} \times \frac{4}{5} = \frac{12}{10} = 1\frac{2}{10} = 1\frac{1}{5}$$

Proportions and ratios

Although proportion and ratio are related they are not the same thing – see example below for clarification.

Example: In a class there are 15 girls and 10 boys. The **ratio of girls to boys is** 15:10, or 3:2, (divide both 15 and 10 by 5) and the **proportion of girls in the class** is 15 out of 25, $\frac{15}{25}$ which simplifies to $\frac{3}{5}$

Questions based on proportions and ratios

Example 1

In a class of 27 pupils, 9 go home for lunch. What is the proportion of pupils in this class that have lunch at school?

Since 9 out of 27 pupils go home, this means 18 pupils have lunch at school.

As a proportion this is 18 out of 27 or $\frac{18}{27}$ which simplifies to $\frac{2}{3}$

Example 2: In a certain work place the ratio of males to females is 2: 3 There are 250 workers altogether. How many of these are male?

Step 1: Find out the total number of parts. You can do this by adding up the ratio parts together. E.g. 2:3 means there are (2+3) = 5 parts altogether. This means 1 part = one fifth of 250 workers = 50 workers.

Since the ratio of male to female is 2:3, there are 2X50 males and 3X50 females

The number of males in this workplace =2X50 = 100

Example 3:

$100 is divided in the ratio 1: 4 how much is the bigger part?

The total number of parts that $100 is divided into is 5 (to find the number of parts simply add the numbers in the ratio, which in this case is 1 and 4)

Clearly, 1 part equals $20 ($100 divided by 5), so 4 parts is equal to $80. This is the required bigger part.

Example 4:

$1500 is divided in the ratio of 3 : 5 : 7

Find out how much the smallest part is worth?

Clearly $1500 is divided into a total of 15 Parts

So each part is worth $100 ($1500 divided by 15)

So 3 parts (this is the smallest part) equals $300

Example 5:

Two lengths are in the ratio 3: 5. If the first length is 150m what is the second length?

If the ratio is 3: 5 then the lengths are in the ratio 150: n

We now need to determine n. We can see that 150 is 50 times 3.

So, n (which is the second length) must be 50 times 5, which equals 250m.

Example 6:

As we have seen, sometimes ratios are expressed in ways, which may not be the simplest form. Consider 5:10

(a) You can re-write 5:10 as 1:2 (divide both sides by 5)

(b) 4 : 10 can be re-written as 2 : 5

(c) 8 : 60 can be re-written as 4 : 30 which, simplifies to 2 : 15

(d) 15 : 36 simplifies to 5 : 12 (divide both sides by 3)

Example 7: A team of 10 people can deliver 6000 leaflets in a residential estate in 4 hours. How long does it take 6 people to deliver these leaflets?

Method: 1 person will take 10 times as long or 4X10 = 40hours

This means 6 people will take $40 \div 6 = 20 \div 3 = 6\frac{2}{3}$ hours = 6 hours and 40 minutes. (Since 1/3 of an hour = 20 minutes)

Scales and ratios

Consider that you are reading a map and the scale ratio is 1: 100000

(This means for every one cm on the map the actual distance is 100000 cm or, put another way every one cm on the map, the distance = 1000 m (divide 100000 by 100. To get the result in metres) now, 1000 m = 1km (divide 1000 by 1000 to get 1 since 1km =1000m)

(Scales can also be used in other areas such as architectural drawings)

Number Work Section 1: Practice Questions

(1) Work out $3\frac{1}{3} - 2\frac{2}{7}$

(2) Find the cubic root of 729

(3) Convert the recurring decimal $3\dot{2}\dot{2}$ to a fraction

(4) I buy a coat for £160 and sell it for £200. What is my percentage profit?

(5) Work out $3\frac{1}{4} \div 4\frac{3}{4}$

(6) I buy a mini-ipad at a discount of 20% for £200. What was its original price?

(7) Given that 8km is approximately 5miles. How many kilometres are there in 75 miles?

(8) Work out $(1\frac{1}{6} \times 1\frac{1}{3}) \div 2\frac{2}{3}$

(9) Work out $(1\frac{1}{4} \times 1\frac{3}{8}) \div 2\frac{3}{16}$

(10) In a class of 28 pupils, 8 pupils have extra maths tuition. What is the proportion of pupils in this class that do not have extra maths tuition?

(11) £280 is divided in the ratio 5: 2: 1. Find the largest part.

(12) I put £1500 in a bank deposit account which gives me a return of 3% per annum compound. How much in total will I have in 3 years?

(13) A map that I am using has a scale of 1: 50000. The distance between the two places I am interested in is 8.5cm. What is the actual distance in km?

Answers to Number Work Section 1: Practice Questions

(1) Answer: $1\frac{1}{21}$

(2) Answer: 9

(3) Answer: $\frac{32}{99}$

(4) Answer: 25%

(5) Answer: $\frac{13}{19}$

(6) Answer: £250

(7) Answer: 120 km

(8) Answer: $\frac{7}{12}$

(9) Answer: $\frac{11}{14}$

(10) Answer: $\frac{5}{7}$

(11) Answer: The largest part is £175

(12) Answer: £1639.10

(13) Answer: 4.25 Km

Number Section 2

Indices and Surds

You are probably already familiar with squares, square roots, cubes and cube roots. Powers, Indices/Index Numbers or exponents are simply the power by which a base number is raised. So just as 4^3 (4 cubed) means 4 to the power of 3, these 'powers' as mentioned earlier are also referred to as indices or index numbers. So 5^6 simply means 5 raised to the power of six. So in this case 5^6 means 5×5 5×5×5×5! (5 is called the base number and 6 is the power or index number) It is interesting to note that if you multiply two or more same base numbers with indices for example: $5^6 \times 5^3$ you simply add the indices to get 5^9 (5 to the power 9). Reason: 5^6 means 5×5×5×5×5×5 and 5^3 means 5×5×5 so $5^6 \times 5^3$ = (5×5×5×5×5×5) × (5×5×5)= 5^9

Similarly, for division, you simply subtract the indices. Consider $5^6 \div 5^3$. This means we need to work out $\frac{5 \times 5 \times 5 \times 5 \times 5 \times 5}{5 \times 5 \times 5}$ which cancels down to 5×5×5 or 5^3 So you can see that when dividing the **same** base numbers with indices you simply subtract the indices.

The examples below will help you to consolidate the manipulation of the same base numbers with indices.

Example 1: $7^8 \times 7^4 \times 7^6 = 7^{18}$ (simply add the indices 8 + 4 +6 =18, hence the answer is: 7^{18})

Example 2: $9^{12} \div 9^5 = 9^7$ (simply subtract 5 from 12 to get 7, hence the answer is: 9^7)

Finally, you can also have negative indices which are inverses of the base numbers with the appropriate indices.

Example 1: $5^{-1} = \frac{1}{5}$ (Also called the reciprocal of 5).

Example 2: $6^{-2} = \frac{1}{6^2}$

Example 3: $5^{-6} = \frac{1}{5^6}$

Fractional Indices: Examples: (i) $2^{1/2}$ (2 to the power of $\frac{1}{2}$), (ii) $27^{1/3}$ (27 to the power of $\frac{1}{3}$. (It's worth noting that $2^{1/2}$ is the same as $\sqrt{2}$, $27^{1/3}$ is the same as $\sqrt[3]{3}$ and $8^{2/3}$ means (8 to the power of $\frac{2}{3}$.)

Rational and Irrational Numbers

Numbers can be either rational or irrational

Any number that can be written as p/q is a rational number, where p and q are whole numbers and q is not zero. Basically, the number is well defined and we know or can predict its pattern.

Examples of rational numbers are: $5 = \frac{5}{1}$, $-2 = \frac{-2}{1}$, $\frac{1}{2} = 0.5$, $\frac{2}{5} = 0.4$, $\frac{1}{3} = 0.33333$ (recurring)

$\frac{0}{5} = 0$, $\frac{4}{33} = 0.1212121212.....$

Examples of irrational numbers are: $\pi, \sqrt{2}, \sqrt{3}$ or $5\sqrt{7}$

For square roots and cube roots those with perfect roots are rational whereas others are irrational. So for example $\sqrt{25} = 5$ is rational, $\sqrt[3]{27} = 3$ is rational but as we saw earlier $\sqrt{2}$ is irrational.

For example π or $\sqrt{2}$ do not have a predictable pattern. We can approximate them but not calculate them exactly.

Summary for Indices:

Rules of indices:

(1) $a^m \times a^n = a^{m+n}$

(2) $a^m \div a^n = a^{m-n}$

(3) $(a^m)^n = a^{m \times n}$

(4) $a^0 = 1$

(5) $a^{-1} = \dfrac{1}{a}$

(6) $a^{-m} = \dfrac{1}{a^m}$

Surds

Surds are simply expressions with irrational square roots. There are some useful rules associated with them.

1: $\sqrt{2} \times \sqrt{2} = \sqrt{4} = 2$

2: $\sqrt{3} \times \sqrt{2} = \sqrt{6}$

3: $\dfrac{\sqrt{6}}{\sqrt{2}} = \sqrt{\dfrac{6}{2}} = \sqrt{3}$

4: $(\sqrt{p} + \sqrt{q})^2 = (\sqrt{p} + \sqrt{q}) \times (\sqrt{p} + \sqrt{q}) = p + 2\sqrt{pq} + q$

5. $(p + \sqrt{q})(p - \sqrt{q}) = p^2 + p\sqrt{q} - p\sqrt{q} - q = p^2 - q$

6. $\dfrac{2}{\sqrt{3}}$ (Multiply top and bottom by $\sqrt{3}$) so we have: $\dfrac{2}{\sqrt{3}} \times \dfrac{\sqrt{3}}{\sqrt{3}} = \dfrac{2\sqrt{3}}{3}$

   (we call this rationalising the denominator)

7. $\dfrac{2}{1-\sqrt{3}}$ to simplify this we need to <u>rationalise</u> the denominator.

   To do this we simply multiply top and bottom by $(1 + \sqrt{3})$

   So we have, $\dfrac{2}{1-\sqrt{3}} = \dfrac{2}{1-\sqrt{3}} \times \dfrac{1+\sqrt{3}}{1+\sqrt{3}} = \dfrac{2(1+\sqrt{3})}{1-3} = \dfrac{2(1+\sqrt{3})}{-2} = -(1+\sqrt{3})$

Note: <u>Leaving your answers as surds is quite respectable, since you can't work out the exact answer on a calculator!</u>

Number Work Section 2: Practice Questions

1 Simplify the following

(a) $2^3 \times 2^4$ (b) $p^6 \div p^7$ (c) $(a^m \times a^n) \div a^k$ (d) $2^{\frac{1}{2}} \times 2^{\frac{-3}{2}}$

2 Write down the following square and cubic roots in its simplest form

(a) $\dfrac{1}{\sqrt{64}}$ (b) $\dfrac{\sqrt{4}}{\sqrt{121}}$ (c) $\dfrac{\sqrt{324}}{18}$ (d) $\sqrt[3]{27}$

(e) $\dfrac{36}{\sqrt[3]{729}}$

3 Write $\sqrt{98} - \sqrt{128} + 5\sqrt{72}$ in the form $a\sqrt{2}$

4 Rationalize the following surds:

(a) $\dfrac{\sqrt{3}}{1-\sqrt{3}}$ (b) $\dfrac{1+\sqrt{2}}{1-\sqrt{2}}$ (c) $\dfrac{5+\sqrt{n}}{5-\sqrt{n}}$

Answers to Number Work Section 2

1 (a) 2^7 (b) p^{-1} or $\frac{1}{p}$ (c) a^{m+n-k} (d) 2^{-1}

2 (a) $\frac{1}{8}$ (b) $\frac{2}{11}$ (c) 1   (d) 3

(e) 4

3   $29\sqrt{2}$

4 (a) $\frac{-3}{2} - \frac{\sqrt{3}}{2}$ (b) $-3 - 2\sqrt{2}$ (c) $\frac{25+10\sqrt{n}+n}{25-n}$

Algebra Section 1

Algebra basics

$x(x+y) = x^2 + xy$

$x^2(x + x^2 + y) = x^3 + x^4 + x^2 y$

In general, $a \times a \times a \times a \ldots\ldots(n \text{ times}) = a^n$

You also need to know these algebraic rules for the multiplication and division of positive and negative numbers.

Reminder: Multiplying positive and negative numbers.

$(+) \times (+) = +$ (a plus number times a plus number gives us a plus number)

$(+) \times (-) = -$ (a plus number times a minus number gives us a minus number)

$(-) \times (+) = -$ (a minus number times a plus number gives us a minus number)

$(-) \times (-) = +$ (a minus number times a minus number gives us a plus number)

Dividing positive and negative numbers.

$(+) \div (+) = +$ (a plus number divided by a plus number gives us a plus number)

$(+) \div (-) = -$ (a plus number divided by a minus number gives us a minus number)

$(-) \div (+) = -$ (a minus number divided by a plus number gives us a minus number)

(−) ÷ (−) = + (a minus number divided by a minus number gives us a plus number)

Summary: **For both multiplication and division, like signs give us a plus sign and unlike signs give a minus sign**

Also when adding and subtracting it is worth knowing that **when you add two minus numbers you get a bigger minus number.**

Example 1: −4 − 6 = −10

When you add a plus number and a minus number you get the sign corresponding to the bigger number as shown below:

Example 2: +6 − 9 = −3, whereas, −6+9 = 3

When you subtract a minus from a plus or minus number you need to note the results as shown below:

Example 3: 6 −(− 3) we get 6+3 = 9 (since −(−3) = +3)

Example 4: 7 −(+3) we get 7 − 3 = 4 (since −(+3) = −3)

In this case note that − (−) = +. Also, +(−) = − and −(+) = −.

Simplifying algebraic expressions

Example 1: Simplify $3m^2 + 4y^3 + 4m^2 - 5y^3$

Method: We add and subtract like terms.

Now $3m^2 + 4m^2 = 7m^2$ and $4y^3 - 5y^3 = -y^3$

Hence, $3m^2 + 4y^3 + 4m^2 - 5y^3 = 7m^2 - y^3$

Example 3: Simplify $ax^2 \times a^4 x^3$

Method: Multiply the two expressions and add the indices for similar terms :

This means $ax^2 \times a^4 x^3 = a^5 x^5$

Example 4 : Simplify $\dfrac{y^3}{x^2} \div \dfrac{y^2}{x}$

Method : Using the rules of indices we get : $\dfrac{y^3}{x^2} \div \dfrac{y^2}{x} = \dfrac{y^3}{x^2} \times \dfrac{x}{y^2} = \dfrac{y}{x}$

(**N.B**. In the above example you can of course cancel down to get the same answer)

Multiplying out brackets.

Example 1: Expand and simplify 3(2x +5) +4(2x+7)

Method: Multiply 3 by each term in the first bracket then 4 by each term in the second bracket. The final step is to simplify by collecting up the like terms.

3(2x+5) +4(2x+7) =6x+15+8x+28 =14x + 43

Example 2: Work out (2x+3)(2x+4)

When we have to multiply out two brackets we have to multiply each term in the first bracket by each term in the second bracket. We then simplify the resulting expression as before. An easy way to multiply out two brackets is to use the grid method as shown below:

First put each of the terms of each bracket on the outside grid as shown

×	2x	+3
2x		
+4		

Step2: Multiply each outside term together. So that for example 2x X 2x =$4x^2$. The other results are shown inside the grid.

×	2x	+ 3
2x	$4x^2$	+ 6x
+ 4	8x	+12

After multiplying out the terms, the answer is found by adding all the terms inside the grid and simplifying the resulting expression.

So we have, $4x^2 + 6x + 8x + 12$ (These are all the terms inside the grid)

Finally, $4x^2 + 6x + 8x + 12 = 4x^2 + 14x + 12$

Another example will help consolidate the process:

Multiply out $(2x - 3)(3x + 2)$

Put the terms of each bracket on the outside of the grid as shown

×	2x	−3
3x	$6x^2$	− 9x
+ 2	4x	− 6

Collecting up all the terms inside the grid we have:

$6x^2 - 9x + 4x - 6$

Now simplify, which gives us $6x^2 - 5x - 6$

Another way of expanding brackets

Example 1: Expand $(x + 3)(x + 2)$

(Multiply the first term of the first bracket by the second bracket and then multiply the second term of the first bracket by the second bracket. Finally simplify the expression.)

So $(x + 3)(x + 2) = x(x + 2) + 3(x + 2) = x^2 + 2x + 3x + 6 = x^2 + 5x + 6$

Example 2: Expand $(2x - 1)(x - 2)$

This equals $2x(x - 2) - 1(x - 2) = 2x^2 - 4x - x + 2 = 2x^2 - 5x + 2$

Typical exam questions

Example 3: Expand $(y^2 + x)(y - x^2)$

$= y^2(y - x^2) + x(y - x^2) = y^3 - y^2x^2 + xy - x^3$

Example 4: Expand and simplify the expression $(x^2 + 2x + 1)(2x^2 + 3x - 2)$

The principle is the same. Take each term in the first bracket and multiply it out by the second bracket. Finally simplify as much as you can.

So $(x^2 + 2x + 1)(2x^2 + 3x - 2) = x^2(2x^2 + 3x - 2) + 2x(2x^2 + 3x - 2) + 1(2x^2 + 3x - 2) = 2x^4 + 3x^3 - 2x^2 + 4x^3 + 6x^2 - 4x + 2x^2 + 3x - 2$

This simplifies to: $2x^4 + 7x^3 + 6x^2 - x - 2$

Example 5: Expand and simplify the expression $(x + y)^2(2x + 3y)$

$(x + y)^2(2x + 3y) = (x + y)(x + y)(2x + 3y) = (x^2 + xy + y^2)(2x + 3y)$

$= 2x^3 + 2x^2y + + 2xy^2 + 3x^2y + 3y^2x + 3y^3$

$= 2x^3 + 5x^2y + 5y^2x + 3y^3$

Difference of two squares

Something useful to remember is the __difference of two squares:__

$$p^2 - q^2 = (p+q)(p-q)$$

Since $(p+q)(p-q) = p(p-q) + q(p-q) = p^2 + pq - pq - q^2 = = p^2 - q^2$

So for example: $16x^2 - 9y^2 = (4x - 3y)(4x + 3y)$

Simplifying Algebraic Fractions

Example 1: Simplify $\dfrac{1}{x+3} + \dfrac{2}{3}$

Method: First find the common denominator which is $3(x + 3)$

Then treat it like you were simplifying a fraction

$$\dfrac{1}{x+3} + \dfrac{2}{3} = \dfrac{1\times 3 + 2(x+3)}{3(x+3)} = \dfrac{3+2x+6}{3(x+3)} = \dfrac{9+2x}{3(x+3)}$$

Example 2: Simplify $\dfrac{2}{x-3} - \dfrac{1}{5}$

As before $\dfrac{2}{x-3} - \dfrac{1}{5} = \dfrac{5\times 2 - 1(x-3)}{5(x-3)} = \dfrac{10-x+3}{5(x-3)} = \dfrac{13-x}{5(x-3)}$

Example 3: Simplify $\dfrac{2}{x-2} + \dfrac{3}{x+2}$

Using the method before we get:

$$\dfrac{2}{x-2} + \dfrac{3}{x+2} = \dfrac{2(x+2)+3(x-2)}{(x-2)(x+2)} = \dfrac{2x+4+3x-6}{x^2-4} = \dfrac{5x-2}{x^2-4}$$

<u>Note</u>: $(x+2)(x-2) = x^2 + 2x - 2x - 4 = x^2 - 4$

Example 4: Simplify $\dfrac{x^2+2x-8}{x-2} \div \dfrac{x^2+4x}{x+2}$

Simplifying we get $\dfrac{(x+4)(x-2)}{x-2} \times \dfrac{x+2}{x^2+4x} = \dfrac{\cancel{(x+4)}\cancel{(x-2)}}{\cancel{x-2}} \times \dfrac{x+2}{x\cancel{(x+4)}}$

Cancelling down as shown above we finally get: $\dfrac{x+2}{x}$

Algebra Practice Questions - simplifying Expressions

Simplify the following:

(1) $\dfrac{x^3}{y^2} \div \dfrac{y^3}{x}$

(2) $4(3x-7) - 5(2x+3)$

(3) $(2x-3)(2x+3)$

(4) $(x-1)^2(x+1)$

(5) $(x^2 + 3x - 1)(3x^2 + 1)$

(6) $(y^2 - 2y + 3)(3y^2 + 3y - 1)$

(7) $\dfrac{1}{x-1} + \dfrac{3}{7}$

(8) $\dfrac{3}{x-3} - \dfrac{1}{6}$

(9) $\dfrac{3}{x-5} + \dfrac{2}{x+5}$

Answers to Simplifying Expressions

Simplify:

(1) Answer: $\dfrac{x^4}{y^5}$

(2) Answer: $2x - 43$

(3) Answer: $4x^2 - 9$

(4) Answer: $x^3 - x^2 - x + 1$

(5) Answer: $3x^4 + 9x^3 - 2x^2 + 3x - 1$

(6) Answer: $3y^4 - 3y^3 + 2y^2 + 11y - 3$

(7) $\dfrac{4+3x}{7(x-1)}$

(8) $\dfrac{21-x}{6(x-3)}$

(9) $\dfrac{5(x+1)}{x^2-25}$

Factorising

Example 1: Factorise: $3x^2 - 6xy$

$= 3x(x - 2y)$ (find the common factor which is 3x in this case)

Example 2: Factorise: $3t^2y - 9t^3$

$= 3t^2(y - 3t)$

Example 3: Factorise $x^2 - 2x - 15$

Find two brackets which when multiplied out together gives you the above expression. We find that $x^2 - 2x - 15 = (x - 5)(x + 3)$

Hence the factors are: **(x – 5)(x + 3)**

Note: We will look at factorizing cubic expressions later!

Practice Questions on factorising algebraic expressions:

(1) $4xp^2 - 3x^2p^3$

(2) $5x^3 - 15x^2 + 25x^4$

(3) $8t^3y^2 - 64t^2y^3 + 16t^2y^2$

(4) $y^2 - 2y - 35$

(5) $6y^2 + 3y - 9$

(6) $7y^2 - 6y - 1$

(7) $9y^2 - 1$

(8) $4n^2 + 12n + 9$

(9) $18p^2 - 3pq - q^2$

Answers to factorising expressions:

(1) $xp^2(4 - 3xp)$

(2) $5x^2(x - 3 + 5x^2)$

(3) $8t^2y^2(t - 8y + 2)$

(4) $(y + 5)(y - 7)$

(5) $(2y + 3)(3y - 3)$

(6) $(7y + 1)(y - 1)$

(7) $(3y + 1)(3y - 1)$

(8) $(2n + 3)(2n + 3)$

(9) $(6p + q)(3p - q)$

Algebraic Substitution and Formula

This is the process of substituting numbers for letters and working out value of the corresponding expression. Some examples that will clarify the process.

Example 1: If k=6 and t=8 work out 2(4k–2t) +kt

Substituting the values of k and t we have:

2(4 × 6–2 × 8) + 6 ×8

=2 × (24 – 16) +48 = 2 ×8 +48 =16+48 =64

So 2(4k – 2t) + kt = 64

Example 2: If t=9 and u= 6 work out $3t^2$ -5u

Substituting appropriately we get:

3 ×9^2 - 5 ×6 = 3 × 81–30 =243 – 30 =213

(Notice, we use the BIDMAS rule to work out the square first and then do the multiplication)

So, $3t^2$ -5u =213

Formula

A formula describes the relationship between two or more variables. You have seen some examples above already. Now let us consider some practical examples.

Example:

The formula for converting the temperature from Celsius to Fahrenheit is given by the formula: $F = \frac{9}{5}C + 32$ (where C is the temperature in degrees Centigrade)

If the temperature is 10 degrees Celsius then what is the equivalent temperature in Fahrenheit?

Using the formula $F = \frac{9}{5}C + 32$, and substituting 10 in place of C, we have $F = \frac{9}{5} \times 10 + 32 = \frac{90}{5} + 32 = 18 + 32 = 50$. Hence, 10 degrees centigrade = 50 degrees Fahrenheit

Explanation of working out above: Remember we multiply and divide before adding and subtracting) There are no brackets to worry about. When working out $\frac{9}{5} \times 10 + 32$, multiply 9 by 10 to get 90, divide this by 5 to get 18, finally add 18 and 32 together to get 50

Example 3: Convert 68 degrees Fahrenheit to degrees Celsius. The formula for converting the temperature from Fahrenheit to Celsius is given by:

$C = \frac{5}{9}(F-32)$, So to change 68 degrees Fahrenheit to degrees Celsius we can substitute for F in the formula $C = \frac{5}{9}(F-32)$, $C = \frac{5}{9}(68-32) = \frac{5}{9} \times 36 = 5 \times 4 = 20$. Hence, 68 degrees Fahrenheit = 20 degrees Celsius

Explanation of the working out above: Using BIDMAS we work out the bracket first. This gives us 68-32 = 36. We now divide this by 9 and multiply by 5. Clearly $36 \div 9 = 4$ and finally $5 \times 4 = 20$

Algebraic Proofs

Example 1: Prove that $(3n+1)^2 - (3n-1)^2$ is a multiple of 4.

Proof: $(3n+1)^2 = 9n^2 + 6n + 1$ and $(3n-1)^2 = 9n^2 - 6n + 1$

So $(3n+1)^2 - (3n-1)^2 = 9n^2 + 6n + 1 - (9n^2 - 6n + 1) =$

$9n^2 + 6n + 1 - 9n^2 + 6n - 1 = 12n = 4(3n)$ which is a multiple of 4

Example 2: Prove that when n and m are positive integers that 2n + 1 + 2m + 1 is always even.

Proof: $2n + 1 + 2m + 1 = 2n + 2m + 2 = 2(n+m) + 2$

Let n + m = k so we have $2(n+m) + 2 = 2k + 2 = 2(k+1)$ which is even

Example 3: Prove that $(n+1)^2 - (n-5)^2 = 12(n-2)$

Proof: $(n+1)^2 - (n-5)^2 = n^2 + 2n + 1 - (n^2 - 10n + 25) =$

$n^2 + 2n + 1 - n^2 + 10n - 25 = 12n - 24 = 12(n-2)$

<u>**Remember for all integer values of n:**</u>

<u>**2n is even and 2n + 1 is odd**</u>

Practice Questions on Algebraic Proofs:

(1) Prove that $\frac{1}{4}(2n+1)(n+4) - \frac{1}{4}n(2n+1) = 2n + 1$

(2) Prove that $(n+2)^2 - (n-6)^2 = 16(n-2)$

(3) Prove that $(n+6)^2 - n(n-3)$ is a multiple of 3 for all positive integers n.

(4) Prove that $(n+1)^2 - (n-5)^2 = 12(n-2)$

(5) Prove that $(3n+1)^2 - (3n-1)^2$ is a multiple of 12 for all positive integers n

(6) Prove that $(p+1)^2 - (p-1)^2 + 1$ is always odd for positive integers

(7) Prove that the difference between the squares of any two consecutive even numbers is always a multiple of 4

Answers to Algebraic Proofs:

(1) $\frac{1}{4}(2n+1)(n+4) - \frac{1}{4}n(2n+1)$

$= \frac{1}{4}(2n^2 + 9n + 4) - \frac{1}{4}(2n^2 + n)$

$= \frac{1}{4}(2n^2 + 9n + 4 - 2n^2 - n) = \frac{1}{4}(8n + 4) = \frac{4}{4}(2n+1) = 2n + 1$

Hence $\frac{1}{4}(2n+1)(n+4) - \frac{1}{4}n(2n+1) = 2n + 1$

Better method: $\frac{1}{4}(2n+1)(n+4) - \frac{1}{4}n(2n+1) = \frac{1}{4}(2n+1)(n+4-n) = \frac{1}{4}(2n+1) \times 4 = 2n + 1$

Note: $\frac{1}{4}(2n+1)$ is the common factor.

(2) $(n+2)^2 - (n-6)^2 = n^2 + 4n + 4 - n^2 + 12n - 36$
$= 16n - 32 = 16(n-2)$

Hence $(n+2)^2 - (n-6)^2 = 16(n-2)$

(3) $(n+6)^2 - n(n-3) = n^2 + 12n + 36 - n^2 + 3n$
$= 15n + 36 = 3(5n + 12)$
Hence, $3(5n + 12)$ is a multiple of 3 for all positive integers of n

(4) $(n+1)^2 - (n-5)^2 = n^2 + 2n + 1 - (n^2 - 10n + 25)$
$= n^2 + 2n + 1 - n^2 + 10n - 25 = 12n - 24 = 12(n-2)$

Hence $(n + 1)^2 - (n - 5)^2 = 12(n - 2)$

(5) $(3n + 1)^2 - (3n - 1)^2 = 9n^2 + 6n + 1 - (9n^2 - 6n + 1)$
$= 9n^2 + 6n + 1 - 9n^2 + 6n - 1 = 12n$ (which is a multiple of 12)

(6) $(p + 1)^2 - (p - 1)^2 + 1 = p^2 + 2p + 1 - (p^2 - 2p + 1) + 1 = p^2 + 2p + 1 - p^2 + 2p - 1 + 1 = 4p + 1$ which is odd

(7) Let 2p and 2p + 2 be two consecutive even numbers
Hence $(2p + 2)^2 - (2p)^2 = 4p^2 + 8p + 4 - 4p^2 = 8p + 4 = 4(2p + 1)$ which is a multiple of 4

Changing the subject

Changing the subject of a formula (re-arranging formulas)

Example 1: In the formula $a = bx + c$ make x the subject

Method: Apply the same rules as you would to equations

In this case subtract c from both sides to get $a - c = bx$

Now divide both sides by b to get $\dfrac{a-c}{b} = x$

In other words $x = \dfrac{a-c}{b}$

Example 2: In the formula $\dfrac{ay^2}{b} + m = k$, make y the subject

Method:

Step 1: Subtract m from both sides to get $\dfrac{ay^2}{b} = k - m$

Step 2: Multiply both sides by 'b' to get $\dfrac{ay^2}{\cancel{b}} \times \cancel{b} = b(k - m)$

(The 'b's on the left hand side cancel)

Hence we now have: $ay^2 = b(k - m)$

Step 3: divide both sides by 'a' (to cancel the 'a' on the left hand side)

We now have $y^2 = \dfrac{b(k-m)}{a}$

Step 4: Take the square root of both sides to get $y = \sqrt{\dfrac{b(k-m)}{a}}$

So we finally find that $y = \sqrt{\dfrac{b(k-m)}{a}}$

Example 3: In the formula $\dfrac{a}{1+t^2} = b + c$, make t the subject

Step 1: Multiply both sides by $1 + t^2$ to get $a = (b + c)(1 + t^2)$

Step 2: Divide both sides by $(b + c)$ to get: $\dfrac{a}{b+c} = 1 + t^2$

Step 3: Subtract '1' from both sides to get: $\dfrac{a}{b+c} - 1 = t^2$

Step 4: This simplifies to $\dfrac{a-1(b+c)}{b+c} = t^2$, which simplifies to $\dfrac{a-b-c}{b+c} = t^2$

Step 5: Take the square root of both sides: $\sqrt{\dfrac{a-b-c}{b+c}} = t$

So finally we have $t = \sqrt{\dfrac{a-b-c}{b+c}}$

Practice Questions on Change the Subject

(1) $C = \frac{5}{9}(F - 32)$, make F the subject

(2) $x = \frac{m(s-t)}{x}$, make x the subject

(3) $y^3 = \frac{3(y^3 + t)}{m}$, make y the subject

(4) In the formula $\frac{1}{f} = \frac{1}{u} + \frac{1}{v}$ make v the subject

(5) Given that $s = ut + \frac{1}{2}at^2$, make a the subject

(6) If $\frac{c}{1+x} = m + n$, make x the subject

Answers to Change the subject:

(1) $F = \dfrac{9C}{5} + 32$

(2) $x = +/- \sqrt{m(s-t)}$

(3) $y = \left(\dfrac{3t}{m-3}\right)^{\frac{1}{3}}$

(4) $v = \dfrac{fu}{u-f}$

(5) $a = \dfrac{2(s-ut)}{t^2}$

(6) $x = \dfrac{c-m-n}{m+n}$

Sequences

Arithmetic Sequences

Working out a general formula for an arithmetic sequence:

You can either use the difference method or the general formula to find the nth term of an arithmetical sequence.

Example 1: (a) Find the nth term of the sequence 5, 8, 11, 14, -----
(b) Hence find the 15th term.

Method: The common difference in this case is 3. So we multiply the common difference by n to get 3n. However each term is 2 more than 3n. (a) Hence the nth term is $3n + 2$

(b) To find the 15th term simply substitute n = 15 in 3n +2.
Hence the 15th term is $3 \times 15 + 2 = 45 + 2 = 47$

Example 2: Find the nth term of the sequence 4, 1, -2, -5

Method: This time the common difference is -3. So the nth term is $7 - 3n$

(Since each term is 7 more than -3n so the nth term is $7 - 3n$

This time consider a general arithmetical sequence as shown:

a, a +d, a+2d, a +3d, a+4d, a+5d, a+6d, ... (d is the common difference). We can see that the second term is a+d, the third term is a+2d, the fourth term is a+3d, the fifth term is a+4d or a + (5-1)d, the sixth term is a+5d or a +(6-1)d

The seventh term is a+6d or a + (7- 1)d, so the nth term is a+(n-1)d

You can check to see if this is right by substituting n=1, 2, 3, 4, 5 and so on to the appropriate numbers in the sequence. See example below:

Example 2: Find the nth term of the arithmetical sequence below:

5, 9, 13, 17 ... this is an arithmetical or linear sequence since the numbers go up by the same constant number. We know the nth term is a + (n - 1)d

In this case a=5 (This is the first term) d = 4 (this is the common difference between each successive number). So, the nth term is 5 +(n – 1)× 4 = 5 +4n –4 = 4n +1,

Fractional sequences

Find the 3rd term of the sequence $\frac{n}{3n-1}$

We simply substitute n =3 in the expression above to give us $\frac{3}{3 \times 3 - 1} = \frac{3}{8}$

<u>Limiting value of a sequence</u>

Find the limiting value of $\frac{3n+4}{6n-3}$ as n $\longrightarrow \infty$

Initially we simply divide each term in the numerator and denominator by n to get: $\frac{3+\frac{4}{n}}{6-\frac{3}{n}}$. We now can see that

as n tends to ∞, the expression $\frac{3+\frac{4}{n}}{6-\frac{3}{n}} = \frac{3}{6} = \frac{1}{2}$

Quadratic Sequence

A quadratic sequence is a sequence of numbers in which the second difference between any two consecutive terms is constant.

Consider the following example: 6, 11, 18, 27, 38…

The first difference is calculated by finding the difference between consecutive terms:

So we get 5, 7, 9, 11…..

The second difference is obtained by taking the difference between consecutive first differences:

This time we get 2, 2, 2, ……

We notice that the second differences are all equal to 2. **Any sequence that has the same common second difference is a *quadratic sequence.***

So we now know the first term is n^2 we now need to find the remaining linear sequence to give us 6, 11, 18, 27, 38, …….

We subtract n^2 (n =1, 2, 3, 4, ….) from this sequence: 6 – 1, 11 – 4, 18 – 9, 27 -16, 38 – 25 to give us 5, 7, 9, 11, 13, …… We can see this is a linear sequence 2n + 3.

Hence the quadratic sequence required is $n^2 + 2n + 3$

Practice Questions on Sequences

(1) Find the nth term of the linear sequence
-8, -3, 2, 7, 12, ...

(2) Find the 8th term of the sequence $\dfrac{5n}{2n-6}$

(3) Find the limiting value of $\dfrac{5n-1}{8n+3}$ as n $\longrightarrow \infty$

(4) Find the nth term of the quadratic sequence:
2, 3, 6, 11, 18, ...

(5) Find the nth term of the quadratic sequence:

5, 11, 19, 29, 41.......

(6) Find the limiting value of $\dfrac{4n+1}{7n+2}$ as n $\longrightarrow \infty$

(7) Find the 10th term of the sequence $\dfrac{7n}{22n+1}$

Answers to Questions on Sequences:

(1)   5n – 13

(2)   4

(3) $\dfrac{5}{8}$

(4) $n^2 - 2n + 3$

(5) $n^2 + 3n + 1$

(6) $\dfrac{4}{7}$

(7) $\dfrac{70}{221}$

Algebra Section 2

Solving equations

Example 1: Solve the equation $5x - 1 = 2x + 8$

First add 1 to both sides, which gives:

$5x = 2x + 9$

Now subtract 2x from both sides to give $3x = 9$

Finally divide both sides by 3 to get x=3.

(Notice each step simplifies the equation further)

Example 2: Solve the equation $5(2x + 1) = 4(2x + 1)$

To solve this first multiply out the bracket which gives:

$10x + 5 = 8x + 4$

(Multiply each term outside the bracket by each term inside the bracket)

Now subtract 5 from both sides, which gives:

$10x = 8x - 1$

Now subtract 8x from both sides, which gives:

$2x = -1$

Finally, divide both sides by 2 to get x= –1/2 or –0.5

Example 3: Solve the equation $\dfrac{2x}{3} + 5 = 7$

We can simplify this to $\frac{2x}{3} = 2$ (by subtracting 5 from both sides)

Now multiply both sides by 3 to get the expression below:

2x =6 , so x =3

Example 4

Solve the equation $\sqrt{4 - \frac{x+3}{3x+2}} = 3$

Although this might look complicated the basic rule is whatever you do to one side you must do the same to the other.

Step 1: Square both sides so we get $4 - \frac{x+3}{3x+2} = 9$

Step 2: Cross –multiply everything by the denominator (3x + 2)

We get: 4(3x + 2) – (x + 3) = 9(3x +2)

Simplify to get 12x + 8 – x – 3 = 27x + 18

Simplify further to get 11x + 5 = 27x + 18

Subtract 18 from both sides to get 11x – 13 = 27x

Now subtract 11x from both sides to get – 13 = 16x

Which is the same as 16x = -13, this means $x = \frac{-13}{16}$

Solving linear equations with inequalities

Example 1: Solve the inequality 2x +5>9

This simply says 2x + 5 is greater than 9. To find x still use the rules of a simple equation. That is, whatever you do to one side you must do to the other.

If 2x +5>9, then 2x >4 (by taking away 5 from both sides)

Now, divide both sides by 2 to get x >2. Our answer for x is all values greater than 2.

Example 2: Solve the inequality 2(5x − 1) ≥ 3x + 14

Method:

This simplifies to 10x − 2 ≥ 3x + 14

Subtract 3x from both sides to get 7x − 2 ≥ 14

Add 2 to both sides to get 7x ≥ 16

Dividing both sides by 7 to get x ≥ $\frac{16}{7}$ ⟹ x ≥ $2\frac{2}{7}$

Example 3: Solve the inequality 4 − 2x < 16

⟹ -2x < 12 ⟹ -x < 6 ⟹ x > -6

(Note: <u>In inequalities when you divide both sides by -1 you also change the sign of the inequality</u>)

Example 3: Show the inequality -2 < x ≤ 2 by way of a number line.

The answer is shown below:

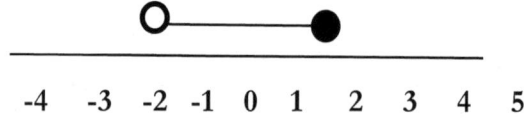

-4 -3 -2 -1 0 1 2 3 4 5

(Note the convention that a dark shaded circle implies 2 is included because x ≤ 2 and at -2 unshaded (open) circle implies -2 is not included because it is <)

Solving Word Problems using Algebra

Examples:

(1) Fatima and Louise have £350 between them. Louise has £80 less than Fatima. How much do they each have?

Method: Let the amount Fatima has be represented by x Hence, Louise has x – 80. We know that the sum of the two amounts = £350. That is x + x – 80 =350. Simplifying, we get 2x – 80 =350. Now add 80 to both sides so we have 2x - 80 +80 = 350 + 80. Which means 2x = 430, or x = 215. This means Fatima has £215 and Louise has £135 (Since Louise has £80 less than Fatima)

(2) The cost of a coat after a 20% discount is £85. What was its original price?

Method: Let the original price be £x. This means x – 20% of x = 85. Or x – 0.2x = 85, which simplifies to 0.8x =85. Now divide both sides by 0.8. So we get x = 85÷0.8 = 106.25. Hence the original price is £106.25

(3) The area of a rectangle is $162m^2$. The length of the rectangle is two times the width. What is the length and width of the rectangle?

Method: Let the width =w, hence the length = 2w. We know that the area of a rectangle is length × width = 2w×w = $2w^2$. The area of the rectangle is given as $162m^2$. Hence, $2w^2 = 162$. Dividing both sides by

2, we get $w^2 = 81$. Hence w = $\sqrt{81}$ =9. So the width is 9m and the length is 18m.

(4) John's annual salary is $\frac{3}{4}$ of Hilary's salary. Hilary's salary is twice Betty's. The total salary between them is $450,000. How much did each of them earn?

Method: Let Hilary's salary be x (in dollars). Hence, John's salary is $\frac{3}{4}$x. Also, since Hilary earns twice as much as Betty, then Betty earns half of Hillary's = $\frac{1}{2}$x. Finally, we know that $x + \frac{3}{4}x + \frac{1}{2}x = $ $450,000, simplifying $2\frac{1}{4}x = 450{,}000$. Or, $\frac{9}{4}x = 450{,}000$. This means 9x = 1,800,000 or x = 200,000. So Hilary earns $200,000, John earns $150,000 (three quarters of Hilary's amount) and Betty earns $100, 000 (half of Hilary's salary)

(5) The sum of two numbers is 30 and the difference between them is 8. What are the two numbers?

Method: Let the unknown numbers be x and y. This means x + y = 30, and x – y = 8, If we add the above two equations we get 2x = 38 (The y's cancel). Hence x = 19 and y = 11

(6) The second number is $\frac{3}{4}$ of the first number. The sum of two numbers is 5.25. What are the two numbers?

Method: The sum of the two numbers is 5.25. Let one of the numbers be x. Hence, $x + \frac{3}{4}x = 5.25$. Simplifying, we get 1.75x = 5.25, dividing both sides by 1.75 we get x = 3. So $\frac{3}{4}x = 0.75 \times 3 = 2.25$. So the two numbers are 3 and 2.25.

Simultaneous Equations

We saw earlier that simple equations allow us to solve problems involving one unknown. When you have to solve problems involving more than one unknown you need more than one equation to solve these.

A simultaneous equation with two variables (meaning two unknowns), say x and y, typically involves two equations. The problem is then to find the values of x and y which satisfies the equations at the same time. Another way of saying simultaneous is 'at the same time'.

We will first consider one traditional method of solving simultaneous equations.

Example 1: Solve the simultaneous pair: $2x + 4y = 5$

$$3x + 5y = 9$$

First let us understand the problem. The problem is to find the values of x and y such that the equations are true.

Method 1:

Consider the substitution method:

We will express x in term of y in the first equation and substitute for x in the second equation. We have:

$2x + 4y = 5$

$3x + 5y = 9$

From the first equation we have $2x = 5 - 4y$ (subtract 4y from both sides)

So $x = 2.5 - 2y$ (we get this by dividing both sides of the previous expression by 2.)

Now substitute $x = 2.5 - 2y$ in the second equation, which gives us:

$3(2.5 - 2y) + 5y = 9$

So, $7.5 - 6y + 5y = 9$

So, $7.5 - y = 9$

So, $7.5 - 9 = y$

Which gives $y = -1.5$

Now we need to find x. We can substitute the value of y in equation 1 to find x.

The first equation is $2x + 4y = 5$

Substituting for y, we get $2x + 4 \times (-1.5) = 5$

which means $2x - 6 = 5$, So $2x = 11$, hence $x = 5.5$

Check

We can check in equation 1 to see if the values we found satisfy the equation.

The first equation is: $2x + 4y = 5$

Substituting, $x = 5.5$ and $y = -1.5$ we get:

$2 \times 5.5 + 4 \times (-1.5) = 11 - 6 = 5$ as required.

Method 2: Eliminate one of the variables

Consider the simultaneous pair of equations as before:

$2x + 4y = 5$ (1)

$3x + 5y = 9$ (2)

Try and make the x or the y terms the same and then add or subtract the equations to eliminate one of the variables

Suppose we make the 'x' term the same. We multiply equation (1) by 3 and equation (2) by 2. The new equations are now shown below:

$6x + 12y = 15$ (3)

$6x + 10y = 18$ (4)

Now subtract (4) from (3) and we get:

$6x - 6x + 12y - 10y = 15 - 18$

$2y = -3$ hence $y = -3/2 = -1.5$

Now substitute $y = -1.5$ in equation (1) and we get:

$2x + 4\times(-1.5) = 5$ which means $2x - 6 = 5$ or $2x = 11$ so $x = 11/2$ or 5.5

Hence as before $x = 5.5$ and $y = -1.5$

There is another method which involves drawing graphs of the two equations and finding the point at which they intersect.

Simultaneous equations with three unknowns:

Solve the simultaneous equations:

$2x + 3y + z = 4$ ……………..(1)

$3x - 3y + 2z = 2$ …………… (2)

$2x + 3y - z = 2$ ……………..(3)

If we add (1) & (2) and adding (2) & (3) we can eliminate y.

So we now have $5x + 3z = 6$ ………..(4)

$5x + z = 4 \ldots\ldots\ldots(5)$

We can now subtract (5) from (4) to get $2z = 2 \Longrightarrow z = 1$

Substitute $z = 1$ in (5) $\Longrightarrow 5x + 1 = 4 \Longrightarrow x = \dfrac{3}{5}$

Finally substitute $z = 1$ and $x = \dfrac{3}{5}$ in (1) to get $\dfrac{6}{5} + 3y + 1 = 4$

$\Longrightarrow 3y = 3 - \dfrac{6}{5} \Longrightarrow 3y = \dfrac{15-6}{5} \Longrightarrow 3y = \dfrac{9}{5} \Longrightarrow y = \dfrac{3}{5}$

Hence the solution to the simultaneous equations with three unknowns in this case are $x = \dfrac{3}{5}, y = \dfrac{3}{5}$ and $z = 1$

Practice Questions Algebra Section 2

(1) Solve the following equations

    (a) $3(2x + 4) = 2(5x - 7)$

    (b) $2x + 7 \geq 2$

    (c) $2(5x - 7) > 6(2x - 8)$

    (d) $\dfrac{1}{3-x} = \dfrac{3}{x+5}$

(2) Solve the following simultaneous equations

    (a) $3x - 2y = 5$,

        $2x + 3y = 7$

    (b) $2x - y = 2$

        $5x + y = 8$

    (c) $x + 2y + z = 1$

        $x + 3y - z = 3$

        $2x - 2y + z = 7$

(3) Solve these word problems algebraically

    (a) John and Elizabeth have £375 between them. John has £85 less than Elizabeth. How much do they each have?

    (b) The sum of two numbers is 35 and the difference between them is 6. What are the two numbers?

(4) Solve the inequality $2(11x - 3) \geq 25x + 17$

(5) Show the inequality $-1 < x \leq 2$ by way of a number line.

(6) Solve the equation $\sqrt{2 - \dfrac{2x+2}{4x+1}} = \sqrt{6}$

Answers to Algebra Section 2

1. (a) $x = 6.5$

   (b) $x \geq -\dfrac{5}{2}$ or $x \geq -2.5$

   (c) $x < 17$

   (d) $x = 1$

2. (a) $x = \dfrac{29}{13}, y = \dfrac{11}{13}$

   (b) $x = = \dfrac{10}{7}, y = = \dfrac{6}{7}$

   (c) $x = \dfrac{46}{13}, y = -\dfrac{8}{13}, z = -\dfrac{17}{13}$

3. (a) John has £145 and Elizabeth £230

   (b) The two numbers are 14.5 and 20.5

4. $x \leq -\dfrac{23}{3}$ or $-7\dfrac{2}{3}$

5. The answer is shown below:

   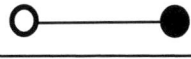

   -4 -3 -2 -1 0 1 2 3 4 5

6. $x = 1$

Algebra Section 3

Solving Quadratic Equations

For the general quadratic equation $ax^2 + bx + c = 0$

The formula for solving the equation is given by: $x = \dfrac{-b \pm \sqrt{b^2 - 4ac}}{2a}$

We can show that this is true by the method of completing the square as shown below.

Consider the general quadratic equation $ax^2 + bx + c = 0$

Dividing through by 'a' we get:

$$x^2 + \frac{b}{a}x + \frac{c}{a} = 0$$

Now we use the method of completing the square

First halve the middle term coefficient and then square the expression on the left hand side as shown below:

$$\left(x + \frac{b}{2a}\right)^2 = x^2 + \frac{b}{a}x + \frac{b^2}{4a^2}$$

Adjusting to get the original expression, we have:

$$\left(x + \frac{b}{2a}\right)^2 - \frac{b^2}{4a^2} + \frac{c}{a} = x^2 + \frac{b}{a}x + \frac{c}{a}$$

We can write $\left(x + \dfrac{b}{2a}\right)^2 - \dfrac{b^2}{4a^2} + \dfrac{c}{a} = 0$

$$\left(x + \frac{b}{2a}\right)^2 = \frac{b^2}{4a^2} - \frac{c}{a}$$

Simplifying the right hand side we get:

$$\left(x + \frac{b}{2a}\right)^2 = \frac{b^2}{4a^2} - \frac{4ac}{4a^2}$$

$$\left(x + \frac{b}{2a}\right)^2 = \frac{b^2 - 4ac}{4a^2}$$

$$x + \frac{b}{2a} = \pm \frac{\sqrt{b^2 - 4ac}}{2a}$$

$$x = -\frac{b}{2a} \pm \frac{\sqrt{b^2 - 4ac}}{2a}$$

$$x = \frac{-b \pm \sqrt{b^2 - 4ac}}{2a}$$

Example:

Solve the equation $2x^2 - 5x + 2 = 0$ using the quadratic formula.

Method:

When the above equation is compared to the general equation

$ax^2 + bx + c = 0$

We can see that a = 2, b= -5 and c =2

Since, $x = \dfrac{-b \pm \sqrt{b^2 - 4ac}}{2a}$

By substituting the above values we can see that:

$$x = \frac{-(-5) \pm \sqrt{(-5)^2 - 4 \times 2 \times 2}}{2 \times 2}$$

$$x = \frac{5 \pm \sqrt{25-16}}{4}$$

$$x = \frac{5 \pm \sqrt{9}}{4}$$

$$x = \frac{5 \pm 3}{4} = \frac{8}{4} \text{ or } \frac{2}{4}$$

Hence $x = 2$ or $\frac{1}{2}$

(Note: The formula method is particularly useful if you find it hard to factorise or if a quadratic expression cannot be factorised)

Solving quadratic equations using factorisation when possible:

Example 1: Solve the equation $x^2 + 5x + 6 = 0$

We can factorise the above quadratic equation as $(x + 3)(x + 2) = 0$

This means either $x + 3 = 0 \implies x = -3$ or $x + 2 = 0 \implies x = -2$

Example 2: Solve the quadratic equation $2x^2 - 5x + 2 = 0$

We can write the above equation as $(2x - 1)(x - 2) = 0$

(You can do this by trial and error with a little bit of common sense)

For example the only way to get $2x^2$ is by having x and 2x in the two brackets. Also the only way to get + 2 as the last term is to have +1 and +2 or -1 and -2. Finally, as the middle term is -5x the factors have to be $(2x - 1)(x - 2)$

So if $(2x - 1)(x - 2) = 0$ this means either $2x - 1 = 0$ or $x - 2 = 0$

If $2x - 1 = 0 \implies 2x = 1$ and $x = \frac{1}{2}$ and if $x - 2 = 0 \implies x = 2$

Hence the solution to the quadratic equation $2x^2 - 5x + 2 = 0$ is

Either $x = \frac{1}{2}$ or $x = 2$

Things to note in quadratic equations and the quadratic formula:

(1) $ax^2 + bx + c = 0$ is a quadratic equation providing 'a' is not zero.
(2) There are two solutions (or roots) to a quadratic equation
(3) The roots are real so long as in the formula $x = \frac{-b \pm \sqrt{b^2 - 4ac}}{2a}$ the bit inside the square root is > 0. The bit inside the square root, that is $b^2 - 4ac$, is called the <u>discriminant</u>. Note if $b^2 - 4ac = 0$, there is only one real root
(4) When $b^2 - 4ac < 0$, then the roots are not real.

Below are examples of equations with two solutions (two roots), one solution (one root) and no solution (no real roots)

(a) The equation $x^2 + 2x - 15 = 0$, has two real roots, $x = -5$ and $x = 3$ as shown in the graph below

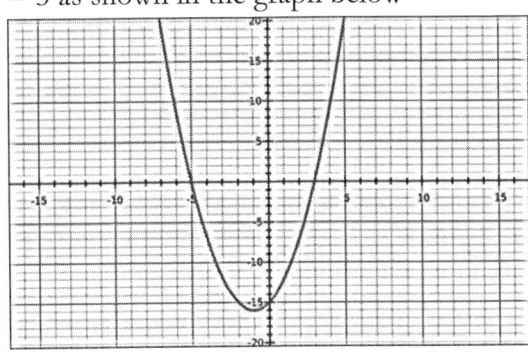

(b) The equation $x^2 - 6x + 9 = 0$ has one real root at x = 3 as shown below

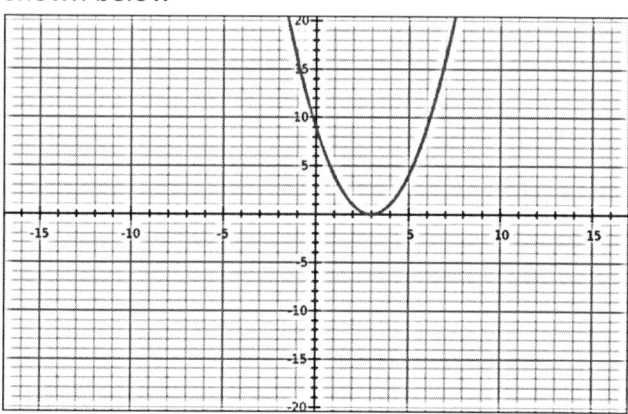

(c) The equation $x^2 - 6x + 12 = 0$ has no real roots as the parabola does not intersect the x-axis at any point.

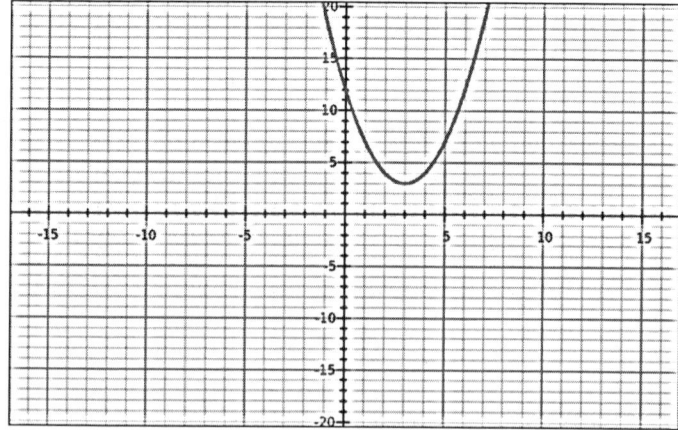

Solving Quadratic Inequalities

Example 1: Solve the quadratic inequality $-x^2 + x + 8 \geq 2$

First re-write this as $-x^2 + x + 6 \geq 0$

Plot its graph and find the values that satisfy this inequality. Namely, values of x, when $y \geq 0$.

First let us see if we can simplify and factorise the equation $-x^2 + x + 6 = 0$. Multiplying through by -1 we get $x^2 - x - 6 = 0$.

$\Rightarrow$ $(x - 3)(x + 2) = 0$ $\Rightarrow$ x = 3 or -2. (Also note that in the initial equation the coefficient of x^2 was negative. This implies that the graph is inverted 'U' shaped as shown below)

y- axis

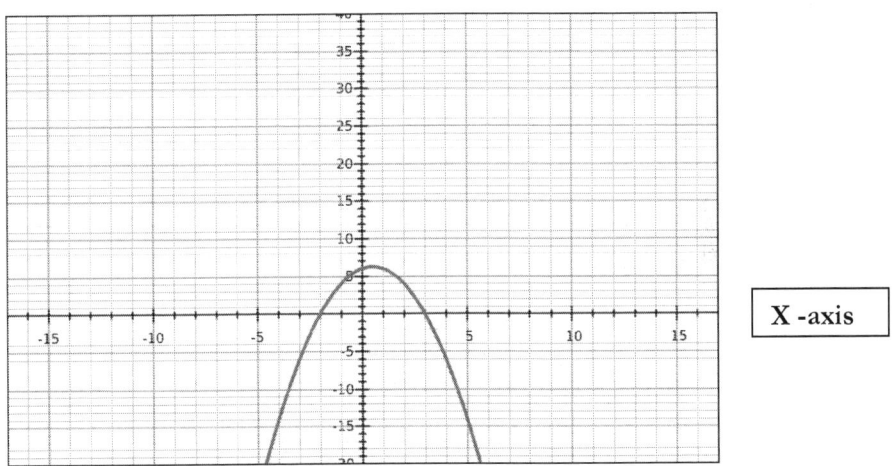

The solution to $-x^2 + x + 8 \geq 2$ is when $-2 \leq x \leq 3$

Example 2:

Solve the inequality $2x^2 + x - 1 > 0$

When it crosses the x-axis we can find the values of x.

That is, $2x^2 + x - 1 = 0 \Rightarrow (2x - 1)(x + 1) = 0$

$\Rightarrow$ $x = \frac{1}{2}$ or $x = -1$. Also since the coefficient of x^2 is positive then the parabola will be **U-shaped as shown below.**

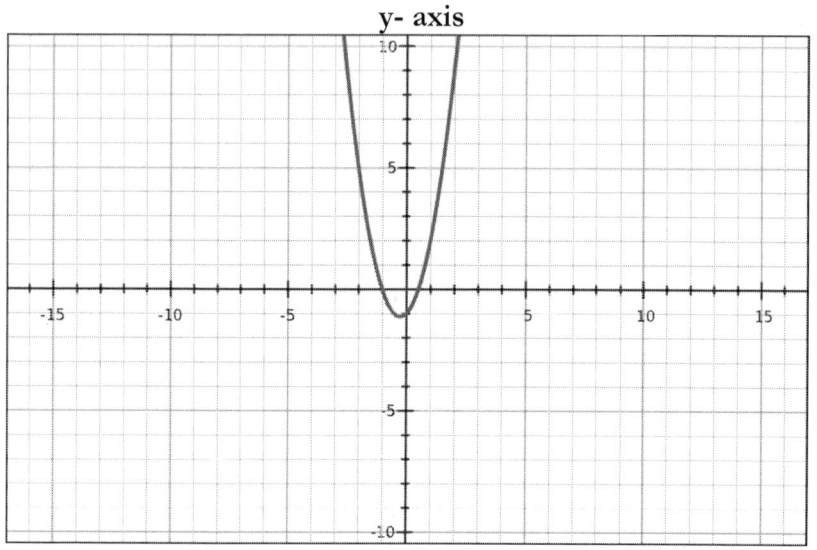

Examining the graph we can see that $2x^2 + x - 1 > 0$ is true when x is less than -1 and when x is greater than $\frac{1}{2}$

That is when $x < -1$ or $x > \frac{1}{2}$

Note: Similar principles have to be applied if you have to solve an quadratic inequality which is < or ≤ rather than > or ≥.

You can best see this visually by actually drawing the graph(s).

Graphs of quadratic equations

Quadratic Equations

These are of the form $f(x) = ax^2 + bx + c$. Note $f(x)$ is called a function of x since y is defined in terms of x.

Consider the example below: Plot the equation $y = 3x^2 - 2x + 1$

First choose suitable values of x say from -3 to + 3 and find the corresponding values of y as shown in the table below:

x	-3	-2	-1	0	1	2	3
$3x^2$	27	12	3	0	3	12	27
-2x	6	4	2	0	-2	-4	-6
+1	1	1	1	1	1	1	1
Y	34	17	6	1	2	9	22

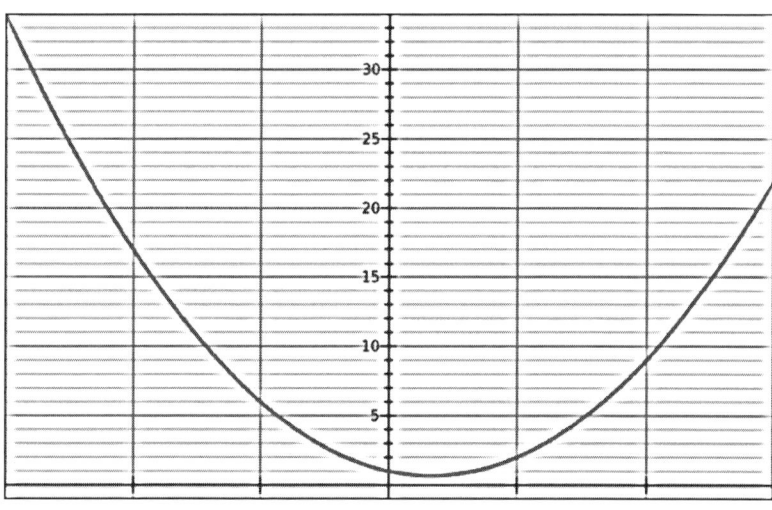

Finding the maximum and minimum values of a parabola (that is the turning points of the graph.

For $f(x) = (x + a)^2 + b$ the turning points are at $x = -a$ and $y = b$

In this case the turning point is a minimum

Example: If $f(x) = (x + 2)^2 + 3$, what are the co-ordinates of the turning points and state whether it is a maximum or a minimum. Also state the line of symmetry.

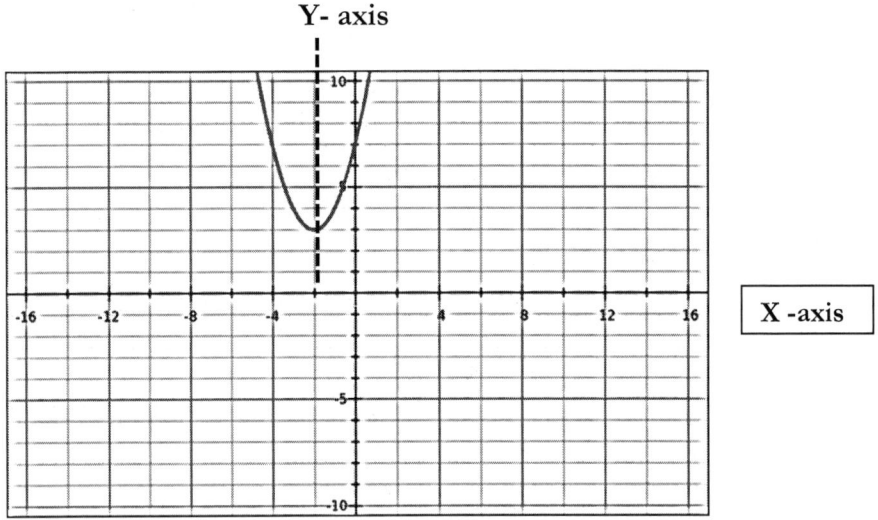

You can see that the graph of $y = (x + 2)^2 + 3$ has a turning point at its minimum -2, 3. The line of symmetry for this function $f(x)$ is at $x = -2$. (Remember $f(x)$ means a function of x, which is the same as $y = (x + 2)^2 + 3$).

Likewise if $f(x) = -(x + a)^2 + b$

Then the turning point is at $x = -a$, $y = b$ and this time is a maximum.

Example 2:

For the equation $y = -(x + 3)^2 + 2$ find the co-ordinates of the turning point and the line of symmetry.

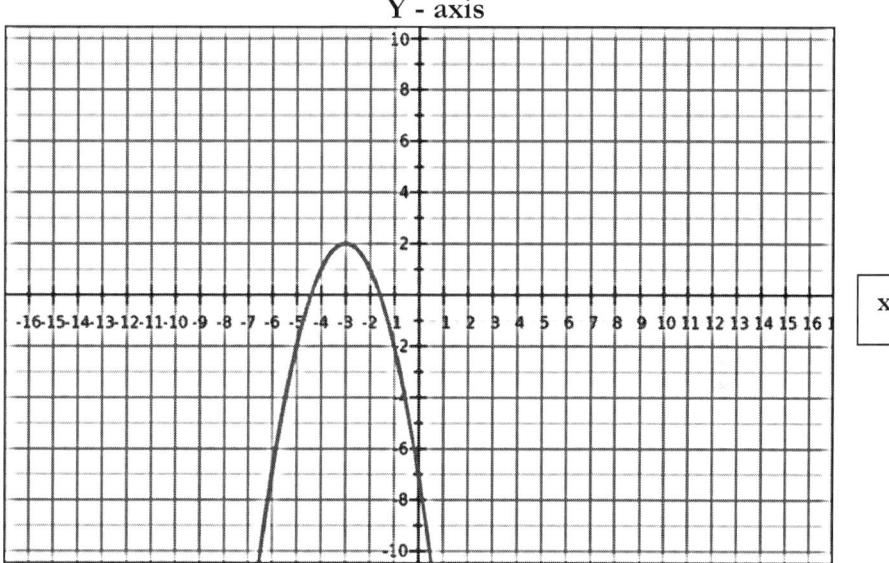

You can see from the graph of $y = -(x + 3)^2 + 2$ above that the turning point is at its maximum where the co-ordinates correspond to (-3, 2). The line of symmetry is at $x = -3$

<u>Note:</u> This means if you are given an equation of a parabola such as $y = x^2 - 2x + 1$, you can re-state it in the form $y = (x + a)^2 + b$, you can then find the co-ordinates of the turning point. In this case the turning point has a minimum value.

Example: Consider the parabola $y = x^2 + 2x + 1$. We can express this as $(x + a)^2 + b = x^2 + 2xa + a^2 + b$. That is the equation of the parabola can be written as $(x + 1)^2 + 0$. Hence the turning point occurs at the point (-1, 0) and it is a minimum.

Cubic equation

Example: Plot the equation $y = x^3 - 1$

x	-3	-2	-1	0	1	2	3
x^3	-27	-8	-1	0	1	8	27
-1	-1	-1	-1	-1	-1	-1	-1
Y	-28	-9	-2	-1	0	7	26

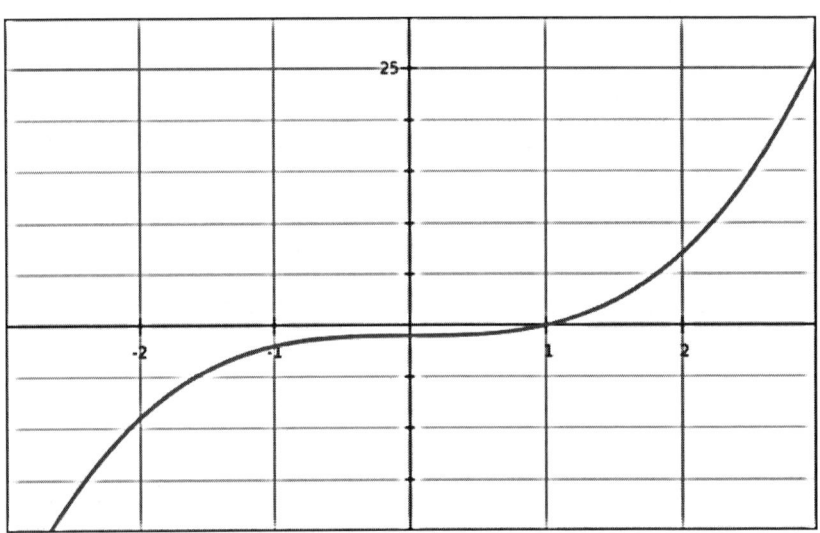

Exponential Graphs

Are of the form y = k to the power of x

Example: To plot the graph of $y = 3^x$ first find some coordinates:

X	-2	-1	0	1	2
y	$3^{-2}=$ 1/9= 0.111	3^{-1} =1/3 =0.333	3^{-0} =$1/3^0$ = $\frac{1}{1} = 1$	$3^1 = 3$	$3^2 = 9$

As we plot these points, you can see that the graph of $y = 3^x$ never falls below the x axis, and when x is positive the y values increase exponentially (or rapidly). See graph below.

Y – axis

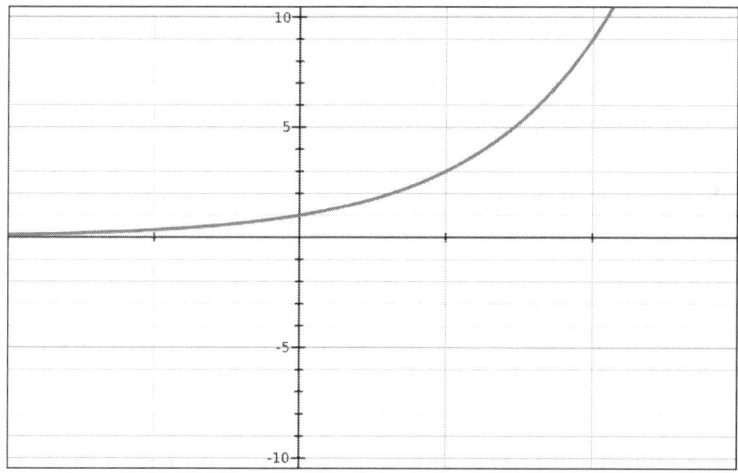

X =axis

Solving equations using graphical methods

Example

You are given that the quadratic equation $y = x^2 - 4x + 8$ and the linear equation $y = 3x - 2$ intersect at two points A and B. Find the co-ordinates of these two points A and B where the equations intersect.

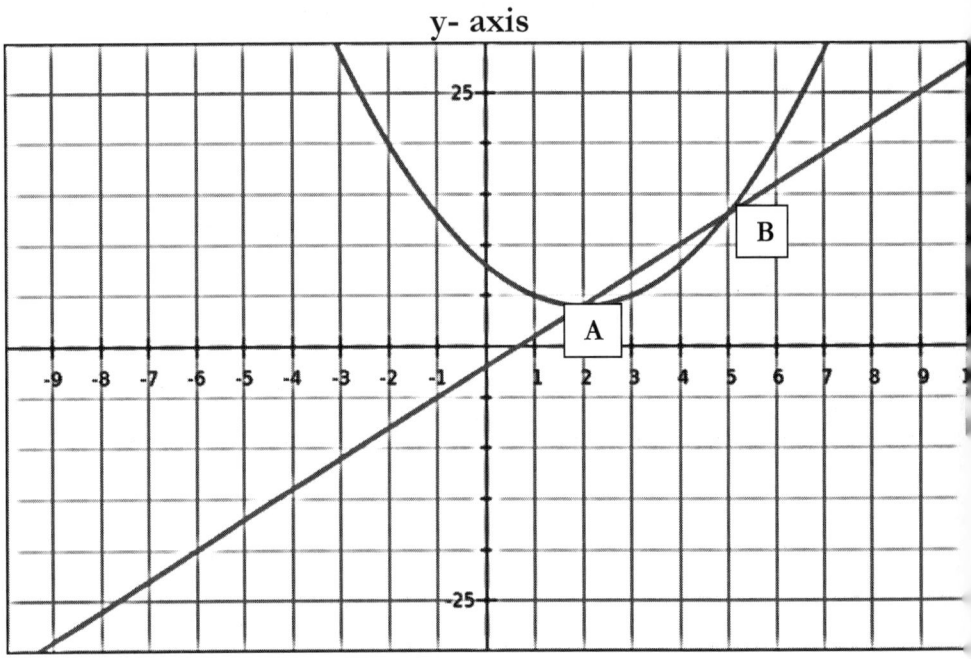

Method: First plot the equations $y = x^2 - 4x + 8$ and the linear equation $y = 3x - 2$ as shown above.

You can see that the co-ordinates of A are (2, 4) and the co-ordinates of B are (5, 13)

Solving equations mathematically when one is linear and the other is quadratic:

The two equations are $y = x^2 - 4x + 8$ and $y = 3x - 2$.

This means $x^2 - 4x + 8 = 3x - 2$

Simplifying this we get $x^2 - 7x + 10 = 0$ ⟹ $(x - 5)(x - 2) = 0$

This means either, x – 5 = 0 or x – 2 = 0 ⟹ x = 5 or 2. We can now find the corresponding values of y by substituting these values in the equation y = 3x – 2

When x =5, y = 3×5 – 2 = 13 and when x = 2, y = 3×2 – 2 = 4

So the co-ordinates of A = (2, 4) and B = (5, 13)

Transformations of functions

When y = f(x +a) the graph moves 'a' units left. (It is the opposite of what you might expect)

Likewise when y = f(x - a) the graph moves 'a' units to the right

Consider the two graphs below (1) f(x) and (2) f(x) = f(x +2)

(1) f(x) = x^2 + 2 and (2) f(x +2) = $(x + 2)^2$ + 2
shown below

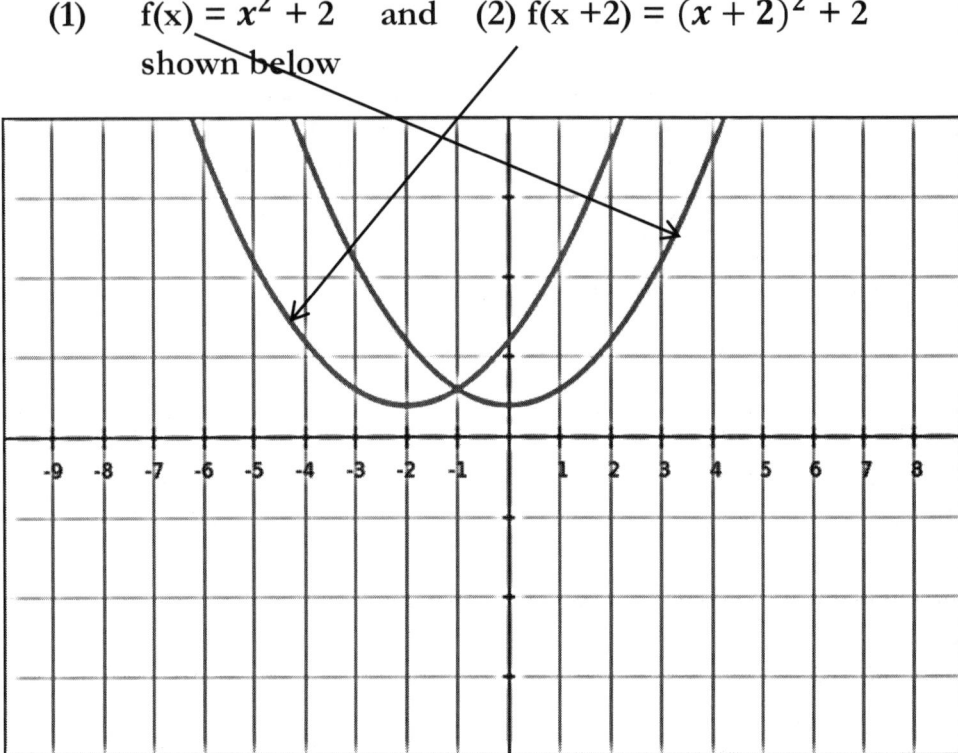

You can see that f(x + 2) has shifted to the left by 2 units (this seems counter-intuitive) but f(x + 2) does shift to the left by 2 units and <u>not</u> to the right.

Other types of transformations involving y = f(x)

It is worth remembering that y = f(x) + a moves up the y-axis by 'a' units and likewise y = f(x) – a moves down the y –axis by 'a' units.

Finally y = k×f(x) or kf(x) simply means the graph of f(x) stretches along the y –axis by a factor of k.

Remainder and Factor Theorem

Remainder Theorem

If you have f(x) and you divide it by x − a the remainder will be f(a)

Example 1: Divide $f(x) = 4x^2 - 4x - 1$ by $g(x) = x - 1$

$$\begin{array}{r} 4x, \text{ remainder } - 1 \\ x-1 \overline{\smash{\big)}\, 4x^2 - 4x - 1} \\ \underline{4x^2 - 4x} \\ 0 + 0 - 1 \end{array}$$

Check: Subsitute x= 1 in f(x) we get 4×1×1 - 4×1 − 1 = -1

Example 2: Divide $f(x) = 2x^3 - 3x^2 - 4x - 1$ by x − 3

Method: To find the remainder just work out f(3)

f(3) = 2×3×3×3 - 3×3×3 - 4×3 − 1 = 54 -27 -12 -1 = 14

Hence the remainder is 14.

Factor Theorem

If f(a) = 0, then this implies there is no remainder and x − a is a factor of the polynomial. Conversely, when x − a is a factor of a given polynomial then f(a) = 0

Example 1: Consider $x^2 - x - 2$ and investigate if x − 2 is a factor.

$f(2) = 2^2 - 2 - 2 = 4 - 2 - 2 = 0 \implies$ x − 2 is a factor.

Example 2: Consider the cubic polynomial $4x^3 - 4x^2 - x + 1$

Show that $2x - 1$, is a factor.

Method: If $2x - 1$ is a factor of $4x^3 - 4x^2 - x + 1$, then $f\left(\frac{1}{2}\right)$ should equal 0.

Let's test this by substituting $x = \frac{1}{2}$ in the cubic equation above.

We get $4\left(\frac{1}{2}\right)^3 - 4\left(\frac{1}{2}\right)^2 - \frac{1}{2} + 1 = \frac{4}{8} - \frac{4}{4} - \frac{1}{2} + 1 = \frac{1}{2} - 1 - \frac{1}{2} + 1 = 0$. Hence $2x - 1$ is a factor of $4x^3 - 4x^2 - x + 1$

The factor theorem can be very useful in finding one of the 'roots' of the equation. That is one possible solution to the equation. We can then try and find other factors by dividing the original cubic equation by $2x - 1$ and factorise the expression we are left with. Although usually you don't have to divide as you can spot the other factors quickly.

See examples below in solving cubic equations.

Solving cubic equations

Example 1:

Solve the cubic equation $3x^3 + 4x^2 - 3x + 2 = 0$, given that one of its factors is x + 2

Step 1: Divide $3x^3 + 4x^2 - 3x + 2$ by x + 2

(We can divide this in the normal long division way as shown below)

$$
\begin{array}{r}
3x^2 - 2x + 1 \\
x + 2 \overline{\smash{\big)}\, 3x^3 + 4x^2 - 3x + 2} \\
\underline{3x^3 + 6x^2 } \\
-2x^2 - 3x + 2 \\
\underline{-2x^2 - 4x } \\
x + 2 \\
\underline{x + 2}
\end{array}
$$

So the cubic equation can be written as (x + 2)($3x^2 - 2x + 1$)

In other words if one of the factors is x + 2, the other factor is $3x^2 - 2x + 1$

Example 2:

Solve the cubic equation $2x^3 + 3x^2 - 3x - 2 = 0$

This time we are not given a factor. In this case it is best to look at the constant term 2 and try x = 1, -1, 2 or -2 as possible solutions. (In other words factors of 2) Let's try x = -1 as a possible solution. If it is a solution then f(-1) should = 0.

Substituting f(-1) we get $2 \times 1^3 + 3 \times 1^2 - 3 \times 1 - 2 = 2 + 3 - 3 - 2 = 0$.

Hence we can re-write $2x^3 + 3x^2 - 3x - 2 = 0$

As $(x - 1)(ax^2 + bx + c) = 0$

Clearly a = 2, since the only way to get $2x^3$ is to multiply x from the first bracket by $2x^2$ in the second bracket. Also we can figure out that c = 2. (Since -1 from the first bracket multiplied by c in the second bracket = -2 which means c = 2.

So far we have $(x - 1)(2x^2 + bx + 2) = 0$. Now multiply the 'x' terms out to get 2x −bx = -3x. This means 5x = bx, hence b =5.

So the factors of $2x^3 + 3x^2 - 3x - 2$ are $(x - 1)(2x^2 + 5x + 2)$

We can now factorise the quadratic expression $2x^2 + 5x + 2$ in the usual way into two brackets. That is $2x^2 + 5x + 2 = (2x +1)((x +2)$.

So finally the cubic equation $2x^3 + 3x^2 - 3x - 2 = 0$ can be written as: $(x - 1)(2x +1)((x +2) = 0 \implies$ x = 1 or x = $-\frac{1}{2}$ or x = -2 (Note: We could have also factorised by dividing $2x^3 + 3x^2 - 3x - 2$ by x − 1 using long division shown earlier)

You can also use the graphical method for solving cubic equations as shown below:

Let y = $2x^3 + 3x^2 - 3x - 2$, that is f(x) = $2x^3 + 3x^2 - 3x - 2$. By choosing suitable values of x we can find corresponding values of y and plot the graph. The graph is shown below:

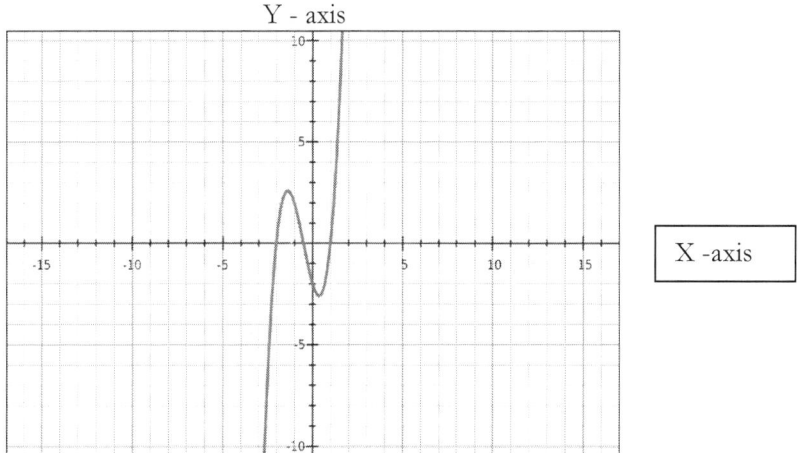

You can see from the graph that the roots are where the curve cuts the x –axis namely at x = –2, –0.5 and 1. Hence the roots of the equation $2x^3 + 3x^2 – 3x – 2$ are when x = –2, or x = –0.5 or x = 1

Practice Questions Algebra Section 3

(1) Solve the quadratic equation $6x^2 + x - 1 = 0$

(2) Find the intersection points where the line $y = x + 2$ meets the curve $y = x^2 - x + 1$

(3) (a) Using the remainder theorem show that when $f(x) = 2x^3 - x^2 - 18x + 12$ is divided by $x - 3$ the remainder is 3

   (b) Prove that the remainder is 3, by dividing $f(x)$ by $x - 3$ using long division

   (c) If $g(x) = (x - 1)(2x^3 - x^2 - 18x + 12)$ work out $g(-2)$

   (d) Evaluate $fg(1)$

(4) A straight line given by the equation $y = 2x + 2$ intersects a curve $y = 2x^2 + 3x - 1$ at two points P and Q.

(a) Find the co-ordinates of these two points.

(b) Work out whether the curve $y = 2x^2 + 3x - 1$ has a maximum or minimum turning point.

(5) A parabola is given by the equation $y = x^2 + 2x - 8$. Write this equation in the form $y = (x + a)^2 + b$ and find the values of 'a' and 'b'.

Answers to Algebra Section 3

(1) Answer: $x = \dfrac{1}{3}$ or $x = \dfrac{-1}{2}$

Method: Factorise $6x^2 + x - 1 = 0$, to get $(3x - 1)(2x + 1) = 0$

⟹ $3x = 1$ or $2x = -1$ ⟹ $x = \dfrac{1}{3}$ or $x = \dfrac{-1}{2}$

(2) Answer: $x = 1 + \sqrt{2}$ and $x = 1 - \sqrt{2}$ and $y = 3 + \sqrt{2}$ and $3 - \sqrt{2}$

Method: Solve the simultaneous equations (linear and quadratic)

⟹ $x + 2 = x^2 - x + 1$ ⟹ $x^2 - 2x - 1 = 0$

⟹ Using the quadratic formula: $x = \dfrac{-b \pm \sqrt{b^2 - 4ac}}{2a}$

⟹ $x = \dfrac{2 \pm \sqrt{2 \times 2 - 4 \times 1 \times (-1)}}{2 \times 1} = \dfrac{2 \pm \sqrt{4+4}}{2} = \dfrac{2 \pm \sqrt{8}}{2} = \dfrac{2 \pm 2\sqrt{2}}{2}$

$= 1 \pm \sqrt{2}$, then substitute for x to find corresponding values of y

(3) (a) Answer: 3

Method: Substitute $x = 3$ in $f(x)$ ⟹ $f(x) = 2 \times 27 - 9 - 54 + 12 = 3$

(b) Answer: See long division below

Method: Do normal long division

$$\begin{array}{r} 2x^2 + 5x - 3 \text{ remainder } 3 \\ x-3 \overline{\smash{\big)}\, 2x^3 - x^2 - 18x + 12} \\ \underline{2x^3 - 6x^2} \\ 5x^2 - 18x \\ \underline{5x^2 - 15x} \\ -3x + 12 \\ \underline{-3x + 9} \\ 3 \end{array}$$

(c) Answer = -84

Method: work out g(-2) by substituting x = -2 in equation g(x)

⟹ (-2 -1)(2×-8 – 4 + 36 + 12) = -3(-16 -4 +48) = -3(28). **Answer = -84**

(d) Answer fg(1)) = 12

Method: First work out g(1). This =0 ⟹ f(0) = 12 Hence fg(1) = 12

(4) (a) Answer: (-1.5, -1) and (1, 4)

Method: Solve the two equations above:

$\implies 2x^2 + 3x - 1 = 2x + 2 \implies 2x^2 + x - 3 = 0$

$\implies (2x + 3)(x - 1) = 0 \implies x = -1.5 \text{ or } 1$

Substituting for x in the equation $y = 2x + 2$, we can find the co-ordinates of the two intersecting points.

(b) **Answer:** Minimum
  Method: Sketch the graph roughly and you will see it is a 'U' shaped curve with a minimum

(5) a = 1 and b = -9

Geometry

Some basic reminders

Reminder of angles, triangles, parallel lines, common shapes & polygons:

(1) Angles on a straight line add up to 180 degrees

(2) The sum of the angles in a triangle add up to 180 degrees

(3) Right angled triangle – one angle in a triangle is 90°

(4) Equilateral Triangle - All sides and angles are equal. Each angle = 60°

(5) Isosceles triangle – base angles are equal and two sides are equal as shown below

(6) Exterior angle of a triangle

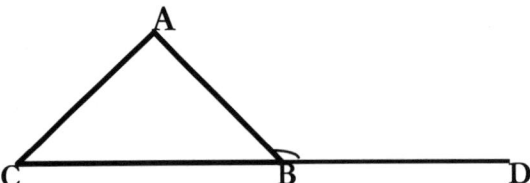

Exterior angle ABD = the sum of the opposite interior angles (Angle ABD = angle BAC plus angle ACB)

Parallel lines

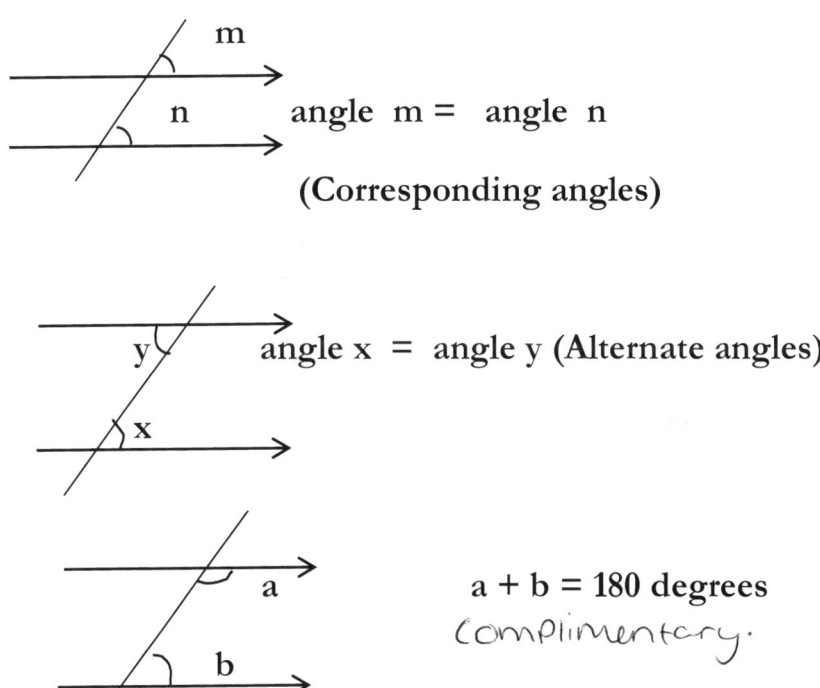

angle m = angle n

(Corresponding angles)

angle x = angle y (Alternate angles)

a + b = 180 degrees
Complimentary.

Quadrilateral – 4 sided shape

The sum of the angles of a quadrilateral add up to 360°

ParallelogramTrapezium

Rhombus (a squashed square)

Kite

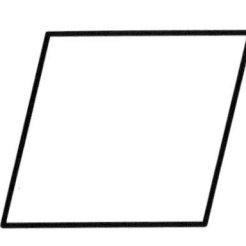
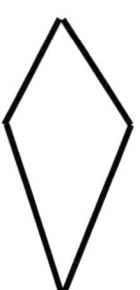

Regular Polygons

A regular polygon is where all the sides and angles are equal.

Consider these typical regular polygons:

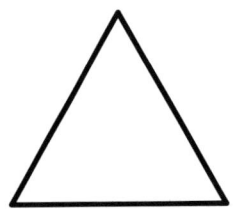

Regular 3 –sided polygon

Equilateral triangle (all sides equal)

Regular 4 – sided polygon

Square

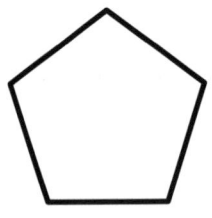

Regular Pentagon (5 sides)

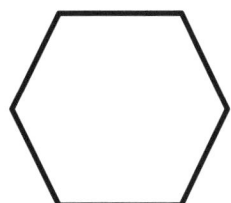
Regular Hexagon (6 sides)

Regular Heptagon (7 sides)

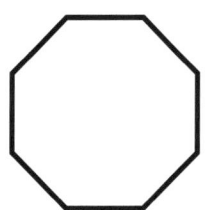
Regular Octagon (8 sides)

Useful formulas for polygons that you should memorize:

Exterior angle $= \dfrac{360}{n}$

Interior angle $= 180° -$ exterior angle $= 180 - \dfrac{360}{n}$ which simplifies to $\dfrac{180n-360}{n} = \dfrac{180(n-2)}{n}$

Angle based examples concerning triangles, parallelograms and polygons

Example 1: Find the interior angle x and the exterior angle y in the triangle below:

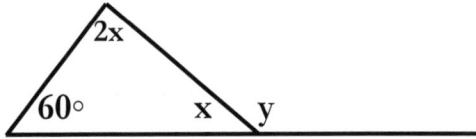

Method: 2x + x + 60 = 180 (interior angles of a triangle add up to 180°). This means 3x + 60 = 180, simplifying this we get 3x = 120. This means x = 40°.

To find y we can use the fact that x + y = 180. This means 40 + y = 180, hence y = 140°. So x = 40° and y = 140°

Example 2:

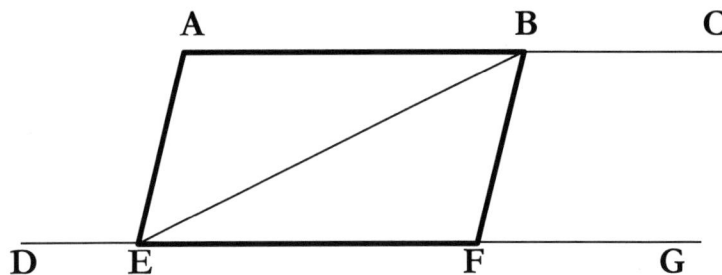

ABFE is a parallelogram. Angle ABE = 30° and angle BFG = 70°

Calculate angle AEB.

Method: Clearly angle BFE = 180 – 70 = 110° Also angle BEF = 30° (Alternate angle to angle ABE).

Now consider triangle BFE. Since we know two angles (angle BEF and angle BFE) we can work out the value of angle EBF. Angle EBF = 180 – (110 + 30) = 40°

Finally, angle AEB = angle EBF (alternate angles)

So angle AEB = 40°

Example 3: Find the interior and exterior angle of a regular octagon.

Using the formula for a regular polygon the interior angle is

$$\frac{180(n-2)}{n} = \frac{180(8-2)}{8} = \frac{180 \times 6}{8} = \frac{90 \times 6}{4} = \frac{90 \times 3}{2} = \frac{45 \times 3}{1} = 135°$$

Pythagoras' theorem

All you need to remember for this is the formula as shown below.

(Note: This theorem is only true for right angled triangles)

The square of the hypotenuse = the sum of the squares of the other two sides.

$h^2 = a^2 + b^2$

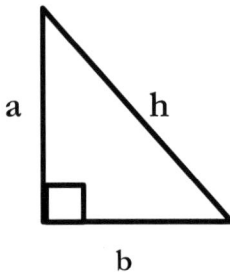

Example 1:

In the triangle below calculate the value of the side b.

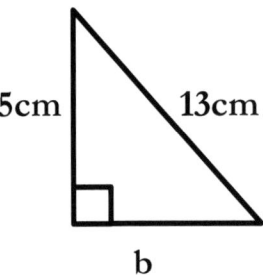

Method: Using Pythagoras' theorem we have $h^2 = a^2 + b^2$

Substituting the values we have $13^2 = 5^2 + b^2$

$\Longrightarrow \ 169 = 25 + b^2 \ \Longrightarrow \ 169 - 25 = b^2 \ \Longrightarrow \ 144 = b^2$

Hence $b = \sqrt{144} = 12$

Example 2:

You are given that in the diagram below: HG =12 cm, GE =8cm, BE = 6cm. Find HB. Give your answer to 4 S.F.

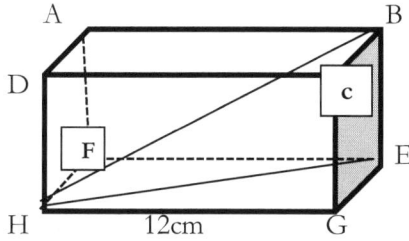

Method: First find HE using Pythagoras' theorem $HE^2 = GE^2 + HG^2$ (Since we know HG & GE)

$\Longrightarrow \quad HE^2 = 12^2 + 8^2 = 144 + 64 = 208 \quad \Longrightarrow \quad HE = \sqrt{208}$

We can now find HB, since $HB^2 = HE^2 + BE^2 \Longrightarrow HB^2 = 208 + 36 = 244$

$\Longrightarrow \quad HB = \sqrt{244} = 15.62$ cm

Circle Theorem

Firstly a few reminders:

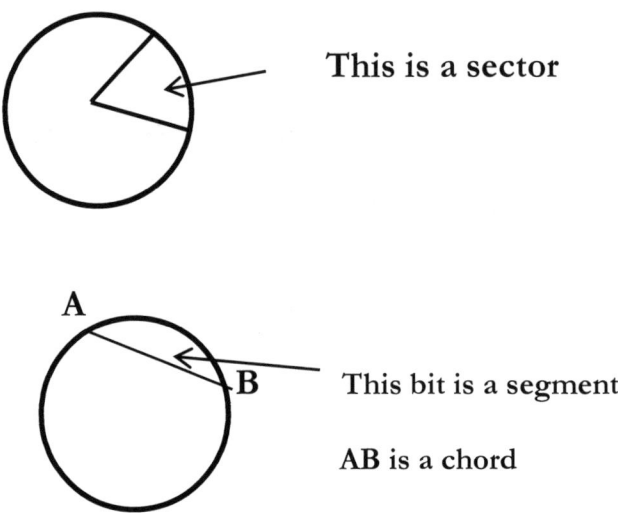

Basic rules of the circle theorem

(1) **Angle subtended at the centre = 2 times the angle at the circumference.**

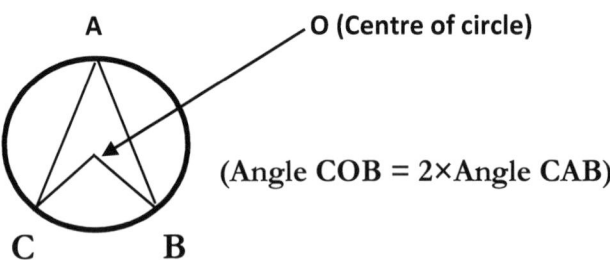

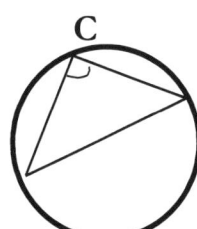

(2) **Angle in a semi-circle is 90°**

(Angle C = 90°)

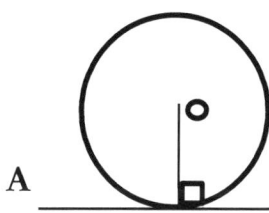

(3) **Radius meets tangent at right angles**

Tangent touches the circle at only one point

(4) The angle between the tangent and chord = angle in the alternate segment

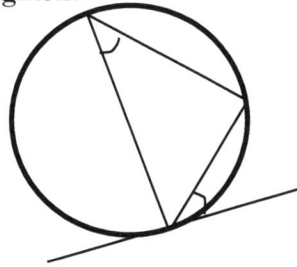

(The two angles shown are equal)

Example 1: Find the angle at the circumference denoted by A if the angle at the centre at O = 120°

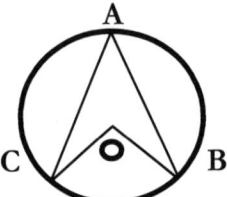

Method: Using the rule that the angle at the centre = 2 times the angle at the circumference, we can work out that the angle at A = half of angle COB. This means that the angle at A = 60°

Example 2: If angle at B = (x + y)°, what is the angle at A?

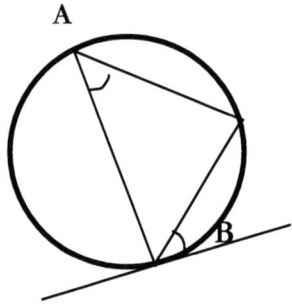

Method: Since the angle between the tangent and the chord (angle B) is equal to the angle in the alternate segment (angle A) then it follows that angle A also = (x + y)°

Area of a Sector of a circle

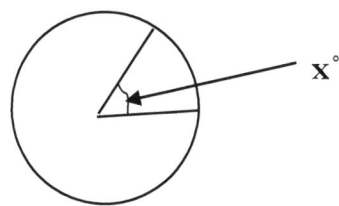

$$\text{Area} = \frac{x}{360} \times \pi r^2$$

Example: Find the area of a sector whose angle at the centre is 60° and the radius of the circle is 10cm. Write the answer to two decimal places

Method:

$A = \frac{x}{360} \times \pi r^2$

$A = \frac{60}{360} \times \pi r^2 \quad A = 0.167 \times \pi r^2$

$A = 0.167 \times 3.142 \times 100 \implies$ **A = 52.5 cm^2**

Equation of a Circle

If a circle has radius 'r' and centre (p, q) then its equation is given by:

$$(x - p)^2 + (y - q)^2 = r^2$$

If the circle has its centre at the origin (0, 0) and its radius is '1' then its graph is as shown below. In this case the corresponding equation is $x^2 + y^2 = 1$

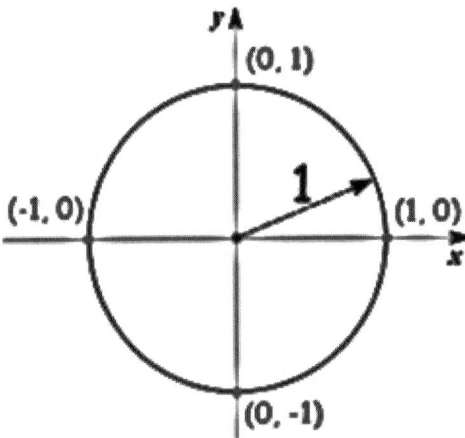

Example 1:

Find the equation of a circle whose radius is 5 and its centre is (2, 3)

Method: Using the formula above the equation of the circle is:

$$(x - 2)^2 + (y - 3)^2 = 5^2$$

We can re-write this as $x^2 - 4x + 4 + y^2 - 6y + 9 = 25$. Simplifying this we get:

$x^2 - 4x + y^2 - 4x - 6y + 13 = 25 \implies x^2 - 4x + y^2 - 4x - 6y - 12 = 0$

Example 2:

Find the centre of a circle and its radius if its equation is $(x - 3)^2 + (y + 4)^2 = 36$. Clearly the centre is 3, -4 and the radius is $\sqrt{36} = 6$
Method: Using the formula for the equation of a circle $(x - p)^2 + (y - q)^2 = r^2$ (where p, q are the centre co-ordinates and r is the radius)

Example 3:

Find the centre of a circle and its radius if the equation is $x^2 - 2x + y^2 - 4y - 11 = 0$

Method: Although this looks a bit tricky we need to make this equation similar to: $(x - p)^2 + (y - q)^2 = r^2$ first.

If we expand this we get $x^2 - 2px + p^2 + y^2 - 2yq + q^2 = r^2$

Equating this equation with the one given we get $-2p = -2$ which means $p = 1$

Likewise for the y term we get $-2q = -4$ which means $q = 2$

So the equation is $(x - 1)^2 + (y - 2)^2 = r^2$

Expanding the brackets we get $x^2 - 2x + 1 + y^2 - 4y + 4 - r^2 = 0$

$\implies x^2 - 2x + y^2 - 4y + 5 - 16 = 0$ (we put -16 to adjust to make the equation the same as the original given. $\implies (x - 1)^2 + (y - 2)^2 = 16$
Hence r = 4 and the centre of the circle is 1, 2

Practice Questions on equation of a circle

Find the equation of the circles (1 – 4) below:

(1) Centre (0, 2) radius 3

(2) Centre (1, -5) radius $\sqrt{2}$

(3) Centre (8, 15) radius 15

(4) Find the centre of a circle and its radius if its equation is $(x-5)^2 + (y+3)^2 = 49$

(5) Find the centre of a circle and its radius if the equation is $x^2 - 2x + y^2 - 4y - 44 = 0$

(6) PQ is a diameter of a circle. P is (-3, 6) and Q is (5, 12). Find the equation of the circle.

(7) The equation of a circle is given by $(x-3)^2 + (y-4)^2 = 49$

   (a) Find the centre of the circle as well as the length of its diameter

   (b) Sketch the circle on a graph

Answers to equation of circle questions

(1) $(x-0)^2 + (y-2)^2 = 9$ which simplifies to: $(x)^2 + (y-2)^2 = 9$

We can re-write this as $x^2 + (y^2 - 2y + 4) = 7 \implies x^2 + y^2 + 2y + 4 = 9 \implies x^2 + y^2 + y - 5 = 0$

(2) $(x-1)^2 + (y+5)^2 = 2 \implies x^2 + y^2 - 2x + 10y + 64 = 0$

(3) $(x-8)^2 + (y-15)^2 = 225 \implies x^2 + y^2 - 16x - 30y + 64 = 0$

(4) Clearly the centre is 5, -3 and the radius is $\sqrt{49} = 7$
Method used: Using the formula for the equation of a circle $(x-p)^2 + (y-q)^2 = r^2$ (where p, q are the centre co-ordinates and r is the radius)

(5) The equation is $x^2 - 2x + y^2 - 4y - 44 = 0$
Method:
Step 1: Equation of a circle can be represented as $(x-p)^2 + (y-q)^2 = r^2$
Step 2: If we expand this we get $x^2 - 2px + p^2 + y^2 - 2yq + q^2 = r^2$
Step 3: Equating this equation with the one given we get – 2p = -2 which means p = 1. Likewise for the y term we get -2q = -4 which means q = 2. So the equation of the circle is $(x-1)^2 + (y-2)^2 = 49$

(6) Equation of the circle is $(x-1)^2 + (y-9)^2 = 25$

Method: First find the centre of the circle from the points of the diameter given. Clearly this is the mid-point of PQ = $\frac{-3+5}{2}, \frac{6+12}{2}$ = (1, 9). Now we need to find the length of the radius using Pythagoras's theorem. The length is $\sqrt{(1-5)^2 + (6+12)^2}$ = $\sqrt{25}$ = 5. So the equation of the circle is: $(x-1)^2 + (y-9)^2 = 25$

(7) Answer: Centre is (3, 4) and diameter 14 units

(7) **Method:**
The equation of a circle is given by $(x - a)^2 + (y - b)^2 = r^2$
Where the centre is (a, b) and radius is r units. This means in the equation given the centre is (3, 4) and radius 7 or diameter 14 units.

(b) Answer is shown by the sketch below

Method: We know the centre is at (3, 4) and the radius is 7 units hence the

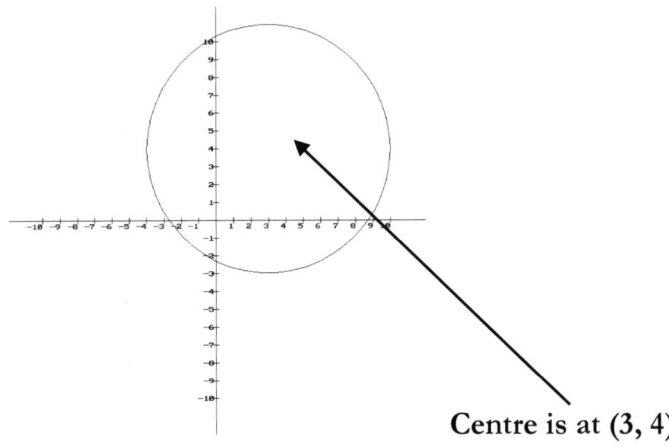

Centre is at (3, 4)

Areas and Volumes of common shapes

Perimeters, Areas and Volumes of common shapes

Consider the shapes below:

(1) Rectangle

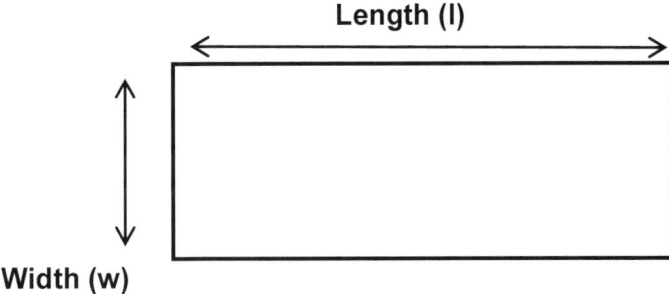

Area of a rectangle = Length X Width or l × w

Perimeter of a rectangle = 2l + 2w (distance around the rectangle)

Note: Area is measured in units squared, e.g. cm^2 or m^2 and perimeter (distance all round a shape) is measured in the appropriate units e.g. cm or m

(2) Triangle

Area of a Triangle = 1/2 × base × height or $\frac{b \times h}{2}$ (The height is the perpendicular height relative to the base)

When you don't know the height of a triangle you need to know that the area of a triangle can be worked out using the formula: $\frac{1}{2}ab\sin C$

Example: Find the area of the triangle shown below:

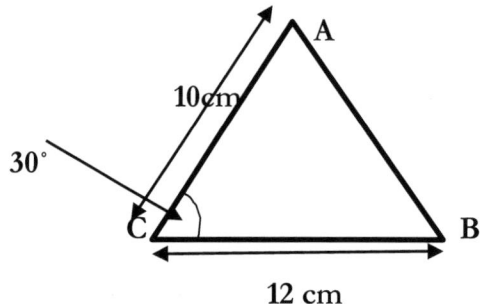

Using the formula Area = $\frac{1}{2}ab\sin C$ ➡

Area = = $\frac{1}{2} \times 12 \times 10 \times \sin 30 = 60 \times 0.5 = 30 \ cm^2$

Area of a Trapezium

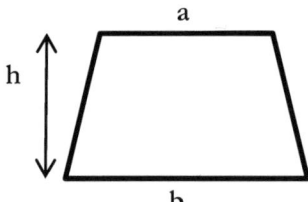

Area of a trapezium

Is equal to half the sum of the parallel sides × perpendicular height) = $\frac{(a+b)h}{2}$

Area of a circle is πr^2 (this means the value of π(pi) multiplied by radius squared)

Circumference of a circle (distance all the way round a circle) = $2\pi r$ or πd.

Note: **Diameter of a circle** = 2 × Radius

Approximate value of π = 3.142

Volume of a cuboid (or a box)

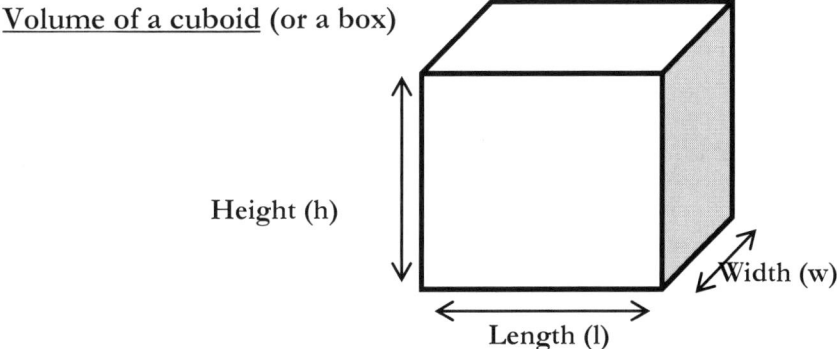

Volume of a cuboid is Height × Length × Width or V = h×l×w (units cubed e.g. cm^3 or m^3, etc)

The surface area of the cube = 2(lw + hw + lh)

You need to also need to know the volume of other 3-D shapes.

For example the volume of a cylinder or any prism is its **base area× height**

The volume of a cylinder $=\pi r^2 h$ (where r is the radius and h is its height)

The surface area of a cylinder $= 2\pi rh + 2\pi r^2$

Example 1: Find the volume of the cylinder shown whose radius is 10cm and height is 15cm

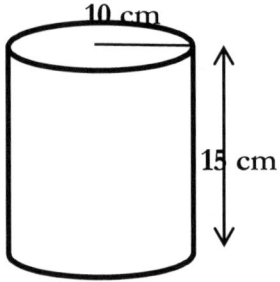

Method: Volume = Base area ×height = area of circle × height = $\pi \times 10 \times 10 \times 15 = 3.142 \times 100 \times 15 = 314.2 \times 15 = 4713 \ cm^3$

Example 2: Find the volume of the triangular prism below. Whose height is 8cm, the width is 6cm and length is 9 cm

(**Note**: A prism is any solid that has a uniform cross – section)

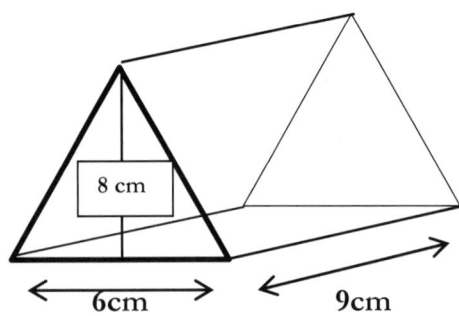

Method: Volume of the triangular prism is its triangle base area × length
= $\frac{1}{2}$×6×8×9 =216 cm^3

Volume of a pyramid or a cone

Volume of a pyramid = $\frac{1}{3}$ base area ×height

Volume of a cone also = $\frac{1}{3}$ base area ×height

Surface Area of a cone = area of circular base + curved area of cone = $\pi r^2 + \pi rl$ (where r is the radius and l is the slant length)

Example 3:

Find the volume of the pyramid shown below. It has a height of 11 metres and a square base whose sides are 6m each.

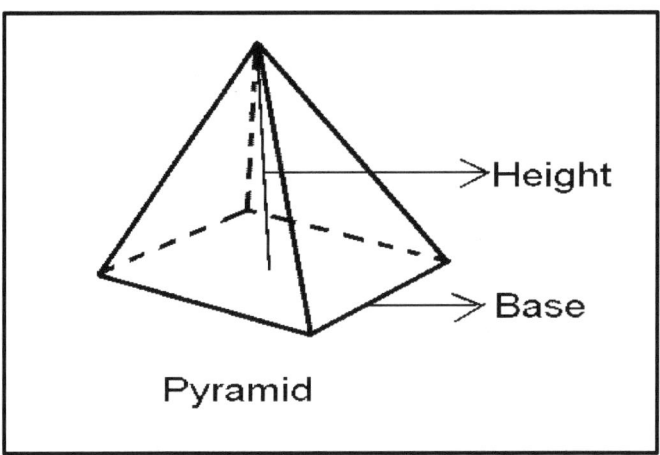

Pyramid

Step 1: Volume of a pyramid = $\frac{1}{3}$ base area ×height

Step 2: Volume = $\frac{1}{3}$ ×6 ×6×11 = 2 ×6 ×11 =12×11 = 132 m^3

Example 4: Find the volume of the cone below whose radius is 5cm and the height is 12 cm. Give your answer to 4 significant figures.

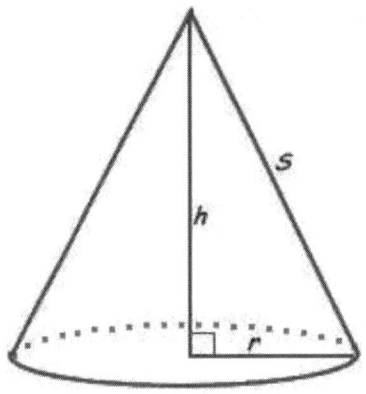

Volume of cone = $\frac{1}{3}\pi r^2 h$

Hence Volume = = $\frac{1}{3} \times \pi \times 5^2 \times 12 = 314.2 \ cm^3$

Another volume question that may come up in the exam is to work out the volume of a sphere:

Volume of a sphere = $\frac{4}{3}\pi r^3$ (where r is the radius)

Example 5:

The volume of a cylinder is $3000 cm^3$. Its radius is 6.5cm. Calculate the height of the cylinder to 3 S.F.

Method: Step 1: Volume of cylinder = $\pi r^2 h$, so $\pi r^2 h = 3000 cm^3$

Step 2: Substitute the value of π and r and make h the subject

$3.142 \times 6.5^2 \times h = 3000$, h = $\frac{3000}{3.142 \times 42.25}$ = $22.59896 \ cm^3$ = $22.6 cm^3$ to 3 S.F.

<u>It is also useful to know the volume and surface area of a sphere.</u>

Volume of a sphere = $\frac{4}{3}\pi r^3$ (where r is the radius)

So volume of a hemisphere = $\frac{2}{3}\pi r^3$

Surface area of a sphere = $4\pi r^2$

Linear equations

These are of the form y = mx + c

where **m** is the **gradient or slope** and c is the value of y when x = 0

Example 1: y = 3x - 1

The graph is shown below.

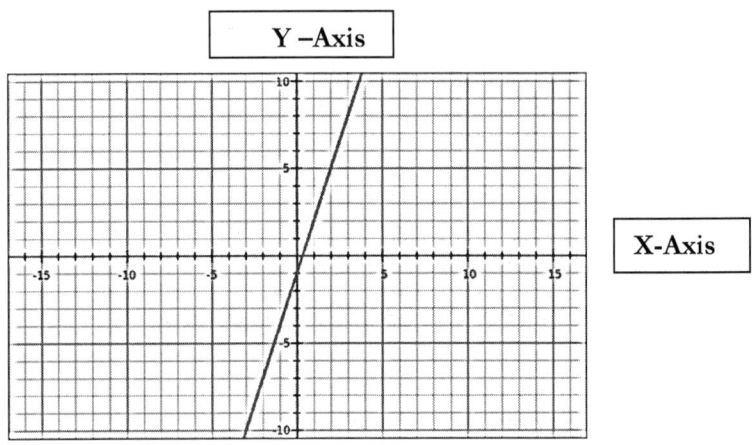

You can see that the equation y = 3x – 1 crosses the y –axis at y = - 1 (this is the called the intercept)

In other words when x = 0, y = 3×0 – 1 = -1

The '3' in the 3x bit refers to the gradient or the slope of the graph.

So in general a linear equation is of the form y = mx + c, where m is the gradient and c is the value of y when x = 0

Example 1

Plot the equation y = 2x - 3 for values of x = -2 to +2 by completing the table below first. The plotted graph is shown below the completed table.

x	-2	-1	0	1	2
2x – 3	-7	-5	-3	-1	1
y	-7	-5	-3	-1	1

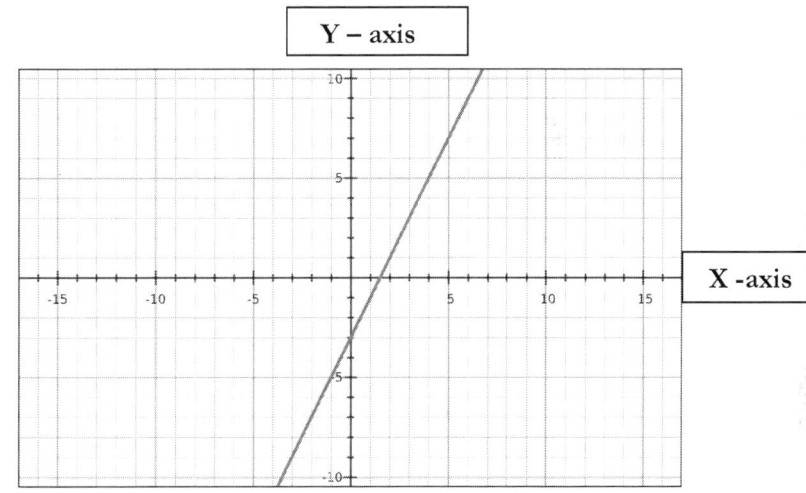

Example 2

The co-ordinates P (2, 3) and Q (4, 6) lie on a straight line.

(i)   Find the mid- point R of the co-ordinates P & Q
(ii)   Find the gradient of the straight line
(iii)   Find the equation of the straight line
(iv)   Find the distance between the two points P & Q

(i) The mid-points are simply the average of the x co-ordinates and the y co-ordinates.

Mid –point for the x co-ordinate = $\frac{x1+x2}{2} = \frac{2+4}{2} = 3$

Similarly, the mid-point of the y co-ordinate = $\frac{y1+y2}{2} = \frac{3+6}{2} = 4.5$

Hence R, the mid-point of PQ is (3, 4.5)

(ii) The gradient of two points that lie on a straight line = $\frac{Difference\ in\ y\ co-ordinates}{Difference\ in\ x\ co-ordinates} = \frac{6-3}{4-2} = \frac{3}{2} = 1.5$

(iii) (a) The equation of a straight line is given by y =mx + c

Since m = 1.5, then the equation is y = 1.5x + c
We also know that it goes through P, Q and R. So we can choose any of these to find the value of C. Let us choose P(2, 3) so the equation is now:
3 = 1.5×2 + C, Hence 3 = 3 + C so in this case C =0
So the equation is y = 1.5x

(b) You can also find the equation of a straight line if you know the co-ordinates of a point it goes through and its gradient by using the formula: y – y1 = m(x – x1). (where x1, y1 are the co-ordinates of the point the line goes through and m is the gradient).

(iv) To find the distance between P & Q we simply use Pythagoras's theorem and find that:

PQ = $\sqrt{(4-2)^2 + (6-3)^2} = \sqrt{4+9} = \sqrt{13}$ units

Using Ratios to find Co-ordinates

Example: You are given that M(-2, 2) and N(3, 12) are end points of a line. Also that P lies on this line such that MP: PN = 3:2. Find the co-ordinates of P.

Method:

Step 1: Find the difference in x and y co-ordinates

Difference in x = 3 – (-2) = 5

Difference in y = 12 – 2 = 10

Step 2: We know that P lies on the line MN in the ratio 3:2

This means P is $\frac{3}{5}$ of the way up from M

Step 3: To find P we use this ratio so that x = $\frac{3}{5} \times 5$ = 3 and y = is $\frac{3}{5} \times 10$ =6

Step 4: Finally, to find P we simply add the co-ordinates of M to the co-ordinates above. So P = (-2 + 3, 2 + 6) = (1, 8)

Working out equations of 'Normals' and 'Parallel' lines

If two lines are perpendicular to each other, then their gradients, m1 & m2 when multiplied together = − 1. That is $m_1 m_2 = -1$ (The line that is perpendicular to given line is called a 'normal')

Example 1

Given the equation $y = 3x - 1$. Find the equation of a line that is perpendicular to it and goes through the co-ordinates P(1, -2)

Method: Gradient of the line $y = 3x - 1$ is 3 (since if $y = mx + c$, then m is the gradient). Clearly if $m1 \times m2 = -1$ then $3 \times m2 = -1$. This means $m2 = -\frac{1}{3}$. Finally if this line goes through P (1, -2) then its equation can be found by using the fact that $y - y1 = m(x - x1)$ where x1, y1 are the co-ordinates it goes through. This means the equation of the **normal** or the **line that is perpendicular to** $y = 3x - 1$ is given by $y - (-2) = -\frac{1}{3}(x - 1)$. (We simply substituted the co-ordinates of P(1,-2) in the equation $y = 3x - 1$)

$\implies y + 2 = -\frac{1}{3}(x - 1) \implies 3y + 6 = -x + 1 \implies 3y + x = -5$

$\implies y = -\frac{x}{3} - \frac{5}{3}$ or $y = -\frac{x}{3} - 1\frac{2}{3}$

Alternative method for finding the perpendicular line:

Since the equation of a straight line is given by y =mx + c, then the equation of the perpendicular line is $y = -\frac{1}{3}x + c$

We can find c by substituting P(1, -2) in this equation. Substituting for x and y we get: $-2 = -\frac{1}{3} \times 1 + c$. This means $c = -2 + \frac{1}{3} = -1\frac{2}{3}$; hence $y = -\frac{1}{3}x - 1\frac{2}{3}$

Finding parallel lines

This is more straightforward **since the gradient of parallel lines are the same**.

Example: Find the line that is parallel to $y = \frac{1}{2}x + 3$ which passes through the point (2, -1)

Method: We know that the equation of a line is given by y = mx + c

Since the gradient of this parallel line is the same, the equation of this parallel line is: $2 = \frac{1}{2} \times (-1) + c \implies c = 2 + \frac{1}{2} = 2\frac{1}{2}$. So the equation of the **parallel line** which goes through (2, -1) is $y = \frac{1}{2}x + 2\frac{1}{2}$

Practice questions in Geometry and Co-ordinate Geometry

(1) Prove that the area of the shape below is $\frac{1}{2}m(p+q) + \pi\frac{m^2}{8}$ cm^2

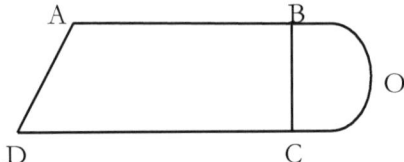

ABCD is a trapezium and BOC is a semi-circle

Where AB = p cm, DC = q cm and BC (the diameter = m cm)

(2) MNOP is a cyclic quadrilateral.

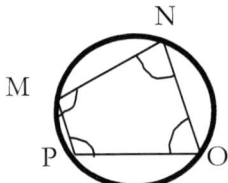

Where angle MNO = n, angle NOP = r, angle OPM = p and angle PMN = m. You are given that p: r: n = 3: 4 : 6. Find the value of angle m.

(3) The equation of a circle is $x^2 + y^2 = 25$ and the point M (3, 4) is on the circle. Work out the equation of the tangent of the circle at M giving your answer in the form y = mx + c

(4) Given the equation y = 3x − 1. Find the equation of a line that is perpendicular to it and goes through the co-ordinates P(1, -2)

(5) You are given that M(-1, 5) and N(4, 10) are end points of a line. Also that P lies on this line such that MP: PN = 1:4. Find the co-ordinates of P.

(6) The equation of a circle C is $(x - 3)^2 + (y - 2)^2 = 13$. Also you are given that K(6, 4) is a point on the circle. (a) Find the radius of the circle (b) Find the equation of the tangent at the point K.

(7) Find the line that is parallel to $y = \frac{1}{4}x - 7$ which passes through the point (2, -1)

Answers

(1) Proof: Area of the shape below is = Area of Trapezium + Area of semi-circle

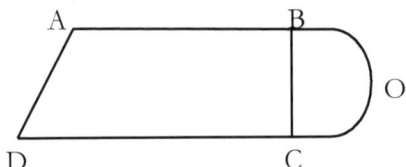

Area of ABCD (trapezium) = $\frac{1}{2}(p+q)m$

Area of the semi-circle BOD = $\frac{1}{2}\pi(\frac{m}{2})^2 = \frac{\pi m^2}{8}$

Hence area of the shape above is $\frac{1}{2}m(p+q) + \pi\frac{m^2}{8}$ cm^2

(2) $m = 100°$

We know that in a cyclic quadrilateral the opposite angles add up to 180°

⟹ $p + n = 180°$

Given the ratios we can say that $3x + 6x = 180°$ ⟹ $9x = 180°$

⟹ $x = 20°$ ⟹ $r = 80°$ ⟹ $m = 180 - 80 = 100°$

(3) $4y + 3x = 25$

Gradient OM = $\frac{4-0}{3-0} = \frac{4}{3}$, hence gradient of tangent at M = $-\frac{3}{4}$

⟹ Equation of tangent = $y - 4 = = -\frac{3}{4}(x - 3)$

Simplifying we get $4y - 16 = -3x + 9$

$\implies 4y + 3x = 25$

(4) Equation is $y = -\dfrac{x}{3} - \dfrac{5}{3}$

Method: Gradient of equation $y = 3x - 1$ is 3. $\implies$ Gradient of line perpendicular to this is $-\dfrac{1}{3}$ $\implies$ Equation of the required line that goes through P(1, -2) is $y - (-2) = -\dfrac{1}{3}(x - 1)$. This simplifies to $y + 2 = -\dfrac{1}{3}(x - 1)$ $\implies 3y + 6 = -x + 1$ $\implies y = -\dfrac{x}{3} - \dfrac{5}{3}$

(5) Q5 Answer: P(0, 5)

Method: Step 1: Find the difference in x and y co-ordinates

Difference in $x = 4 - (-1) = 5$, Difference in $y = 10 - 5 = 5$

Step 2: We know that P lies on the line MN in the ratio 1: 4

This means P is $\dfrac{1}{5}$ of the way up from M

Step 3: To find P we use this ratio so that $x = \dfrac{1}{5} \times 5 = 1$ and $y = $ is $\dfrac{1}{5} \times 5 = 1$

Step 4: Finally, to find P we simply add the co-ordinates of M to the co-ordinates above. So P = (-1 + 1, 4 + 1) = **(0, 5)**

(6) (a) Radius = $\sqrt{13}$

(b) Equation of tangent at point K is $y = -\dfrac{3x}{2} + 26$

Method: Let the centre of the circle be at O, clearly O has the co-ordinates 3,2. Hence gradient of OK is $\dfrac{4-2}{6-3} = \dfrac{2}{3}$

$\implies$ Gradient of tangent at K is $-\dfrac{3}{2}$ (Since OK represents the gradient of the 'normal' to this tangent at K)

⟹ The equation of the tangent at K is $y - 4 = -\frac{3}{2}(x - 6)$

This simplifies to $2y - 8 = -3x + 18$ ⟹ $y = -\frac{3}{2}x + 13$

(7) The equation of the line required is $y = \frac{x}{4} + \frac{3}{2}$

Method: The parallel line has the same gradient as the equation $y = \frac{1}{4}x - 7$ ⟹ gradient of parallel line is $\frac{1}{4}$

⟹ Equation of the parallel line which goes through the point (2, -1) is $y - (-1) = \frac{1}{4}(x - 2)$ ⟹ $y + 1 = \frac{1}{4}(x - 2)$

⟹ $4y + 4 = x - 2$ ⟹ $y = \frac{x}{4} + \frac{3}{2}$

Trig for Right Angled Triangles

Formulae for Right Angled Triangle

In a right-angled triangle you need to know:

The Sine of an angle = the ratio of the Opposite Side to the Hypotenuse

The Cosine of an angle = the ratio of Adjacent Side to Hypotenuse

The Tangent of an angle = the ratio of the Opposite Side to the Adjacent Side

You can also try remembering **SOH-CAH-TOA**

SOH –   SIN (ANGLE) = **OPPOSITE SIDE/HYPOTENUSE**

CAH –   COS (ANGLE) = **ADJACENT SIDE/HYPOTENUSE**

TOA –   TAN (ANGLE) = **OPPOSITE SIDE/ADJACENT**

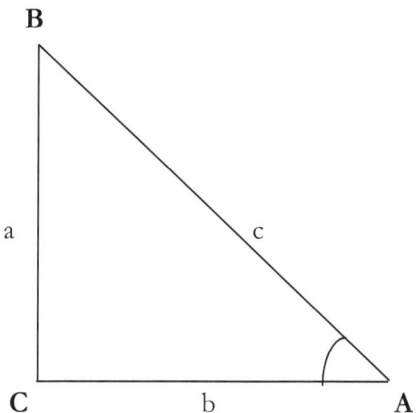

$\text{Sin (A)} = \dfrac{a}{c}$

$\text{Cos (A)} = \dfrac{b}{c}$

Tan (A) = $\dfrac{a}{b}$

Also, using Pythagoras' theorem you can prove that:

$Sin^2 x + Cos^2 x = 1$

Proof: Consider the triangle below:

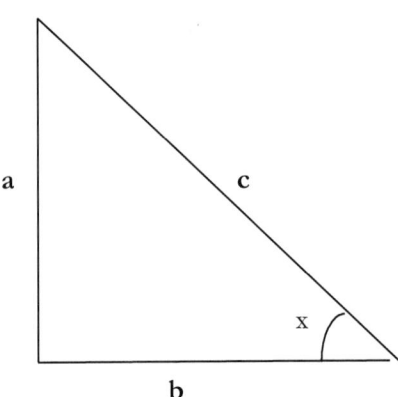

We know that in this triangle

$Sin\ x = \dfrac{a}{c}$ and $Cos\ x = \dfrac{b}{c}$

Hence, $Sin^2 x + Cos^2 x = (\dfrac{a}{c})^2 + (\dfrac{b}{c})^2$

This implies that:

$Sin^2 x + Cos^2 x = \dfrac{(a)^2 + (b)^2}{c^2}$

From Pythagoras' theorem we know that: $a^2 + b^2 = c^2$

Hence, $\sin^2 x + \cos^2 x = \dfrac{c^2}{c^2} = 1$

Trig for non- right angled triangles

Formula for a <u>non-right</u> angled triangle are shown below

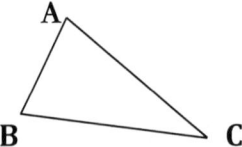

Sine Rule:

$$\frac{a}{SinA} = \frac{b}{SinB} = \frac{c}{SinC}$$

Cosine Rule:

$a^2 = b^2 + c^2 - 2bcCosB$

$b^2 = a^2 + c^2 - 2acCosB$

$c^2 = a^2 + b^2 - 2abCosC$

(Note: Although there are three versions of the formula they all have the same pattern)

Example1: In the triangle below find angle B

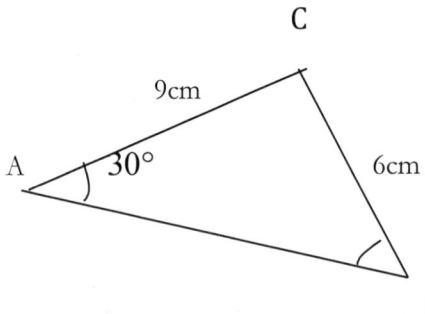

Using the Sine rule: $\dfrac{a}{SinA} = \dfrac{b}{SinB}$ substitute the appropriate known angles and sides in the formula shown

$$\dfrac{6}{Sin30} = \dfrac{9}{SinB}$$

Hence, Sin B = 9 × Sin30/6 = 0.75. So the angle B = **48.6°**

Summary for using the sine rule

You can use the sine rule when: You know two angles and a side.

Also note that the area of a non- right angled triangle is given by $\frac{1}{2}abSinC$

Using the cosine rule

Example: In the triangle shown find the angle C

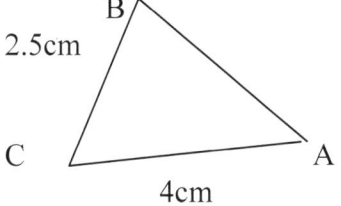

Using the Cosine Rule, we can say that $c^2 = a^2 + b^2 - 2abCosC$

Hence, $CosC = (a^2 + b^2 - c^2)/2ab = (9 + 16 - 6.25)/24 = 18.75/24 = 0.78125$

Hence, C = **38.6°**

Summary for using the cosine rule. (1) You know the lengths of 3 sides or (2) you know two sides and the angle between them

Graphs of Trig Functions

Consider graphs of y = sin(x) and cos(x)

For y=sin(x) maximum value of y = 1 when x = 90° and minimum value of y = -1 when x= -90°. This cycle repeats every 360°

Also note that when x = 0°, sin(0) =0, and when x = 360°, sin(360) = 0

Similarly for y = cos(x). The maximum value of y = 1, when x = 0° and minimum value of y = -1 when x = 180°. Again this cycle repeats every 360°

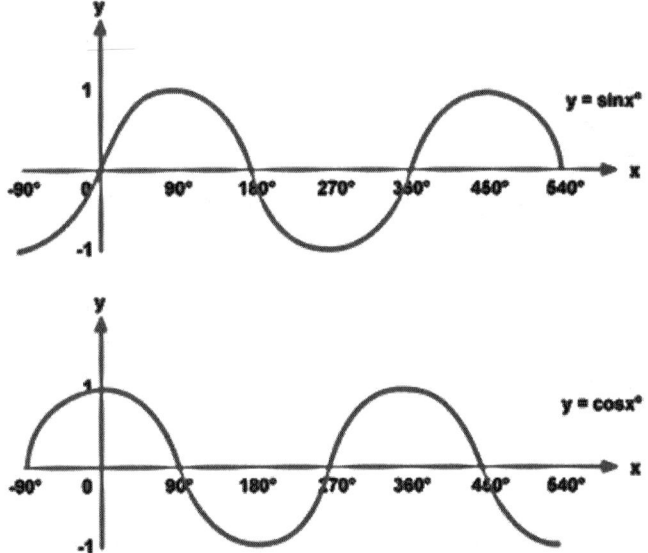

Example of repeating cycles for sin(x)

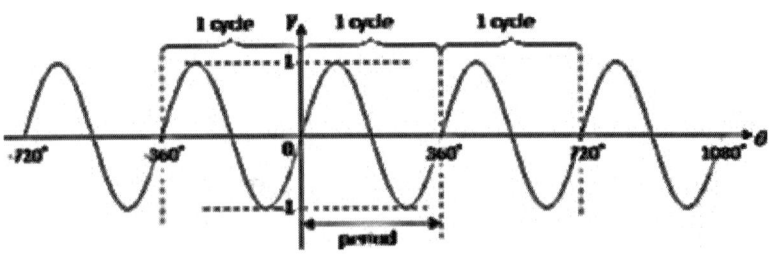

y = tan(x)

The values of tan (x) repeat every 180°. As x approaches 90°, the value of y or tan(x) approaches infinity. The symbol for infinity is ∞.

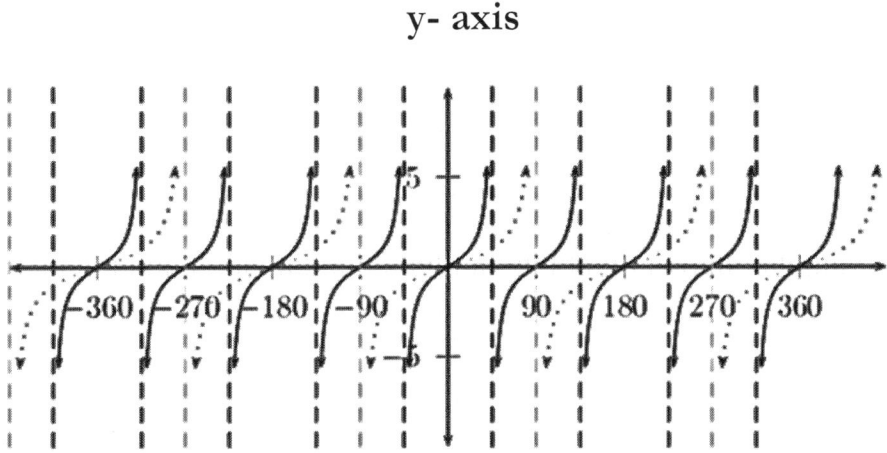

Below is a magnified view of a tangent curve this time of y= tan(θ).

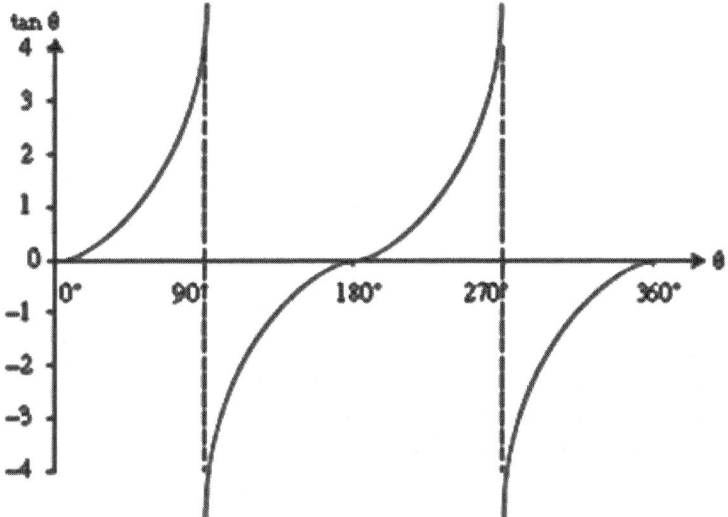

You can see that as θ approaches 90°, tan (θ) tends to infinity

Similarly, as θ approaches -90°, tan (θ) tends to minus infinity

Solving trigonometric equations using graphical methods: Solve the equation sin(x) = 0.5 for values of x between 0 and 360°

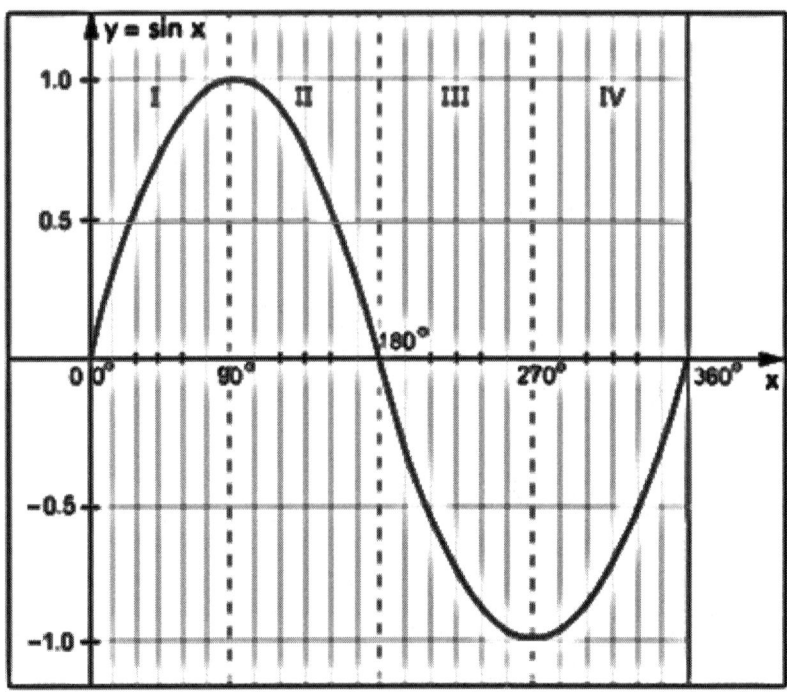

From the graph you can see that when y = 0.5, it meets the curve when x = 30° and 150°

Trig Identities

Rule 1

$sin^2 x + cos^2 x = 1$

Re-arranging, we can also deduce that $sin^2 x = 1 - cos^2 x$

Similarly, $cos^2 x = 1 - sin^2 x$

Rule 2

Tan x = $\frac{sinx}{cosx}$

Example 1: Simplify $sin^2 x + cos^2 x + \frac{sinx}{cosx} - 1$

Method: We know that $sin^2 x + cos^2 x = 1$ and $\frac{sinx}{cosx} = \tan x$

$\implies sin^2 x + cos^2 x + \frac{sinx}{cosx} - 1 = 1 + \tan x - 1 = \tan x$

Example 2: If $\sin x = \frac{1}{2}$, Find the value of cos x in surd form

Method: Using $sin^2 x + cos^2 x = 1$ we can deduce that $\frac{1}{2} \times \frac{1}{2} + cos^2 x = 1$

$\implies \frac{1}{4} + cos^2 x = 1 \implies cos^2 x = \frac{3}{4} \implies \cos x = \frac{\sqrt{3}}{2}$

Example 3: Solve the equation 4sinx + 6 = 9 for values of 0° ≤ x ≤ 90°
(Give your answer to 1 d.p)

Simplifying the equation 4sinx + 6 = 9, we get 4sinx = 3 $\implies \sin x = \frac{3}{4}$
= 0.75. Hence x = 48.6°

Calculus

Differentiation

If y = f(x) then differentiating this function is finding the rate of change of y with respect to x. **It is also the gradient at a point on the curve.** It is represented by $\frac{dy}{dx}$.

If $y = x^n$, then $\frac{dy}{dx} = nx^{n-1}$

Example 1: Differentiate the function $y = 2x^3 + 3x + 7$, and find its gradient when x = 3

Method: First differentiate this function

$\frac{dy}{dx} = 6x^2 + 3 + 0 = 6x^2 + 3$, hence the gradient at x = 3 is 6×3×3 + 3 = 18×3 + 3 = 54 + 3 = 57.

(**Note: when differentiating a constant e.g. any number c we get 0**)

Hence the gradient $\frac{dy}{dx}$ when x = 3 for this function is 57.

Example 2:

You are told that a vehicle travels s metres in t seconds. The formula is given by the equation $s = 4t^2$, find the speed of this vehicle when t = 3 seconds.

Method: $\frac{ds}{dt}$ represents the rate of change of distance with respect to time. This in fact is the speed.

So differentiating, $s = 4t^2$ we get $\frac{ds}{dt} = 8t$. This means the speed of this vehicle at 3 seconds is 8×3 = 24 metres/second.

Stationary points, minima and maxima

For a curve the stationary point occurs when $\frac{dy}{dx} = 0$. (The gradient = 0). To establish whether it is a maximum or minimum you can either sketch the curve near the turning points or find the second derivative. If the second derivative namely, if $\frac{d^2y}{dx^2} > 0$, then this stationary point is a minimum. Also if $\frac{d^2y}{dx^2} < 0$, then the stationary point is a maximum.

Example 3:

You are given that the equation of a curve is $y = 3x^3 - 2x^2 + 2x + 6$

   (a) Find the gradient of this curve when x = 1
   (b) Find the equation of the tangent to this curve at this point
   (c) Find the equation of the normal at this point (x = 1)

(a) **Method:** Gradient = $\frac{dy}{dx} = 9x^2 - 4x + 2$

$\implies$ at x =1 the gradient ($\frac{dy}{dx}$)= 9×1×1 - 4×1 +2 = 9 – 4 +2 = 7

(b) **Method:** When x =1 then the value of y can be found by substituting for x in is $y = 3x^3 - 2x^2 + 2x + 6$ $\implies$ y =3× 1^3 -2× 1^2 + 2×1 + 6 = 3 -2 +2 +6 =9 $\implies$ y = 9. Hence when x = 1, y = 9 and the gradient is 7

Since the equation of a straight line can be found by using y – y1 = m(x – x1)

By substituting the appropriate values we get y – 9 = 7(x – 1)

$\implies$ y – 9 = 7x – 7 $\implies$ y = 7x + 2

(c) **Method:** To find the equation of the normal at this point we first need to find its gradient. Using the fact that m1m2 =- 1 **(Note if m1 is** the gradient of a tangent at a given point and m2 the gradient of the normal at that point then m1×m2 = -1)

We can deduce that 7×m2 = -1 ⟹ m2 = $-\frac{1}{7}$

⟹ The equation of the normal at this point is y − 9 = $-\frac{1}{7}$(x − 1)

Simplifying we get 7y − 63 = -x + 1 ⟹ 7y = 64 − x

Or y = $\frac{1}{7}$(64 − x)

Practice Questions in Trigonometry, Geometry and Calculus

(1) In the triangle below find the length of the side AB. You are given that angle ACB = 32°, AC = 6.2cm and CB = 12.4 cm

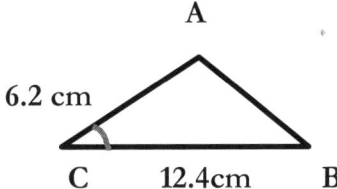

(2) (a) If $\tan(x) = \frac{-5}{3}$ for -90° ≤ x ≤ 90°. Find the value(s) of x in this range.

(b) Show that $\frac{(1+\cos(\theta))(1-\cos(\theta))}{\sin(\theta)} = \sin(\theta)$

(3) A curve is given by the equation $y = 2x^3 + 9.5x^2 + 10x + 8$
(a) Find the stationary points of this curve

(b) Determine if this is a maximum or minimum turning point

(4) (a) Find the tangent to the curve $y = 2x^2 - 2x + 5$ at the point x =1, y =5

(b) Find the equation of the normal to this curve at the same point

(5) (a) Solve the equation $\tan(\theta) = -1$ where $0 \leq \theta \leq 360°$

(b) Solve the equation $3(1 - \sin(x)^2) + 2\sin(x) = 2$ for $0 \leq x \leq 360°$

(6) Differentiate the following:

(a) $y = 3x^3 + 2x$

(b) $y = 2x^2 - 3x + 9$

(c) $y = 5x^{-2} + 4x - 7$

Answers to Practice Questions in Trigonometry and Calculus

(1) Answer: AB = 7.86 cm

Method: in the triangle below use the cosine rule:

$$c^2 = a^2 + b^2 - 2ab\cos C$$

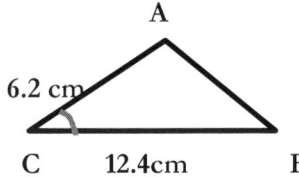

Since $c^2 = a^2 + b^2 - 2ab\cos C$

⇒ $AB^2 = 12.4^2 + 6.2^2 - 2 \times 6.2 \times 12.4 \times \cos 32$

⇒ $AB^2 = 153.76 + 38.44 - 130.4 = 61.8$

⇒ $AB = 7.86$ cm

2 (a) Answer: x = -59.04°

Method: $Tan^{-1}(-5/3) = -59.04°$

(b) See method below

Method: cross multiply by sin(θ) to get :

$(1 + \cos(\theta)(1 - \cos(\theta)) = \sin^2(\theta)$

⇒ $1 - \cos^2(\theta) = \sin^2(\theta)$ ⇒ $1 = \sin^2(\theta) + \cos^2(\theta)$

Since $\sin^2(\theta) + \cos^2(\theta) = 1$ ⇒ $\dfrac{(1+\cos(\theta))(1-\cos(\theta))}{\sin(\theta)} = \sin(\theta)$

3(a) Answer: x = -2.5, y = 11.14

Method: Stationary points occur when $\dfrac{dy}{dx} = 0$

⟹ $\dfrac{dy}{dx} = 6x^2 + 19x + 10 = 0$

⟹ $(3x + 2)(2x + 5) = 0$ ⟹ $x = \dfrac{-2}{3}$ or $x = -2.5$

⟹ Corresponding values of y can be found by substituting the values of x in the equation $y = 2x^3 + 9.5x^2 + 10x + 8$

⟹ When $x = \dfrac{-2}{3}$, y = 4.96 and when x = -2.5, y = 11.14

(b) Answer: when $x = \dfrac{-2}{3}$, turning point is a minimum and when x = 2.5 it is a maximum

Method1: Sketch the curve near the points $x = \dfrac{-2}{3}$, and x = -2.5

Method2: If $\dfrac{d^2y}{dx^2} > 0$, then this turning point is a minimum similarly if

$\dfrac{d^2y}{dx^2} < 0$ then the turning point is a maximum.

$\dfrac{d^2y}{dx^2} = 12x + 19$ ⟹ when $x = \dfrac{-2}{3}$, $\dfrac{d^2y}{dx^2} = 12 \times \dfrac{-2}{3} + 19 = 11$

⟹ $\dfrac{d^2y}{dx^2} > 0$ hence at this point the turning point is a minimum

Also when x = -2.5 then, $\dfrac{d^2y}{dx^2} = 12 \times -2.5 + 19 = -30 + 19 = -11$

⟹ When x = 2.5 the turning point is a maximum

4 (a) **Answer: y = 2x + 3**

Method: At x =1 the gradient $(\frac{dy}{dx})$ = 4x – 2 = 4×1 -2 = 2

Hence the equation of the tangent at (1, 5) is y – 5 = 2(x - 1)

$\implies$ y – 5 = 2x – 2 $\implies$ y = 2x + 3

(b) **Answer: 2y = 11 – x or y = 5.5 - $\frac{x}{2}$**

Method: Gradient of normal can be found using the fact that m1×m2 = -1 (gradient of a line × gradient of its perpendicular = -1)

This means the gradient of the normal is $\frac{-1}{2}$. Hence equation of normal at the same point is y – 5 = $\frac{-1}{2}$ (x – 1)

$\implies$ 2y – 10 = -x + 1 $\implies$ 2y = 11 – x or y = 5.5 - $\frac{x}{2}$

(5) (14) (a) Answer: θ = 135° and 315°

Method: from the graph of y=tan(θ) shown on the next page we can see that there are two values between the range given. Also we can work out θ = $tan^{-1}(-1)$ exactly using the appropriate trig function. Since if tan(θ) = 1 then θ = 45°, by symmetry of the graph the values of θ when tan(θ) = -1 are (180 -45) =**135°** and (360 -45) = **315°**

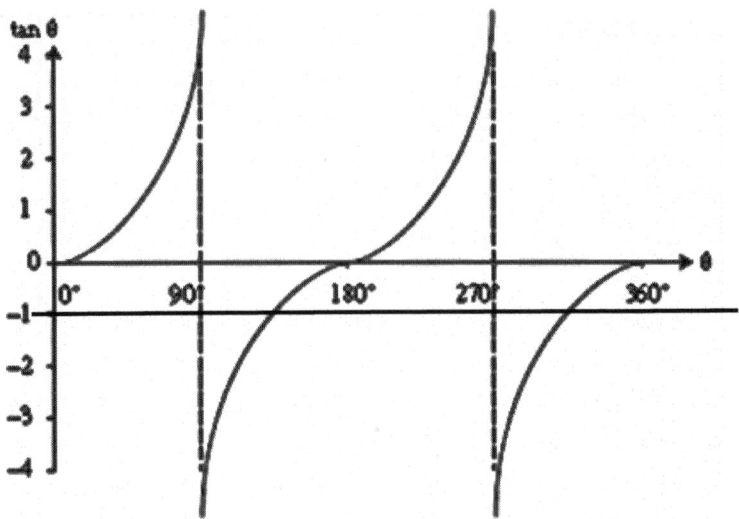

(b) **Answer: x = 90°, 199.47° and 340.53°**

(To solve the equation 3(1 - sin(x)²) + 2sin(x) = 2
for 0 ≤ x ≤ 360° see method below: If you find it easier let
sin(x)= y and then solve the resulting quadratic equation in y.

Method: To solve the equation $3(1 - \sin(x)^2) + 2\sin(x) = 2$ for

$0 \leq x \leq 360°$, Expanding the bracket we have $3 - 3\sin(x)^2 + 2\sin(x) = 2$.

$\implies 3\sin(x)^2 - 2\sin(x) - 1 = 0 \implies (3\sin(x) + 1)(\sin(x) - 1) = 0$

$\implies \sin(x) = -0.33333...$ or $\sin(x) = 1$

From the sine curve (and using trig functions in a calculator as
appropriate) we can see that sin(x) =0.333333, when x = (180
+ 19.47) = **199.47 or** (360 – 19.47) = **340.53°** and sin(x) = 1
when x =90°. Hence the solutions are **90°, 199.47° and 340.53°**

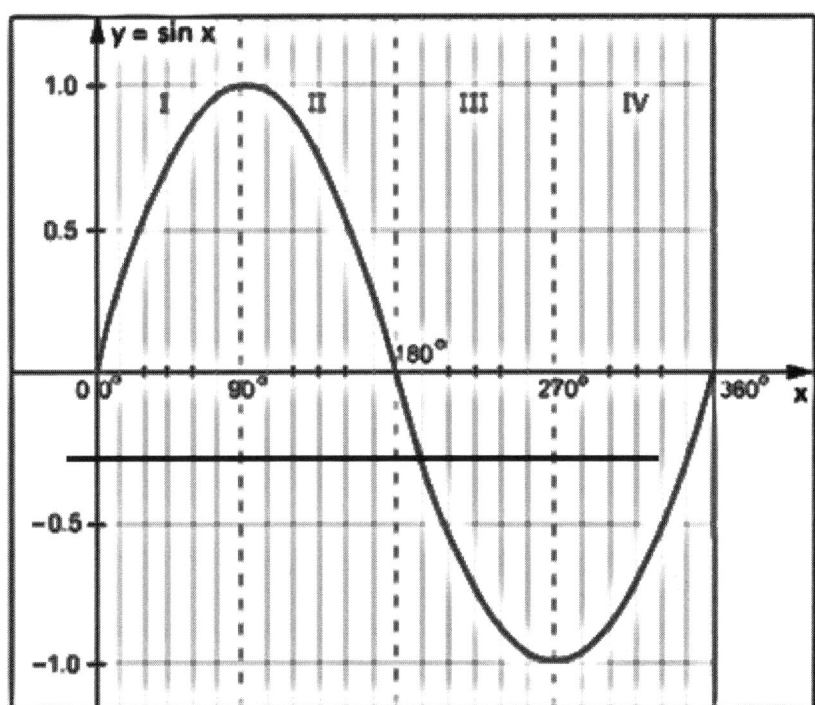

(6) (a) = $9x^2$

(b) = $4x - 3$

(c) = $-10x^{-3} + 4$

(**Method**: In all above cases use the rule if $y=x^n$, then $\frac{dy}{dx} = nx^{n-1}$)

A matrix is simply an array of numbers

As we will see matrices can be very useful for geometrical transformations such as reflections, rotations and enlargements.

Example 1: $\begin{bmatrix} 2 & 1 \\ 3 & 2 \end{bmatrix}$ This is a 2 x 2 matrix. In other words it has two rows and 2 columns

Example 2: $\begin{bmatrix} 1 \\ 2 \end{bmatrix}$ This is a 2 × 1 matrix. It has two rows and one column

So you can see that matrices (plural of matrix) can come in different number of rows and columns. <u>**We need to only worry about these two types of matrices for this exam**</u>

Adding and Subtracting Matrices

You can add or subtract two matrices only if they have the same number of rows and columns.

Example 1: Add matrix **A** and matrix **B** where $A = \begin{bmatrix} 2 & 1 \\ 3 & 2 \end{bmatrix}$ and $B = \begin{bmatrix} 1 & 0 \\ 2 & 1 \end{bmatrix}$

Method: Simply add the elements of each matrix as shown below:

$$A + B = \begin{bmatrix} 2 & 1 \\ 3 & 4 \end{bmatrix} + \begin{bmatrix} 1 & 0 \\ 2 & 1 \end{bmatrix} = \begin{bmatrix} 3 & 1 \\ 5 & 5 \end{bmatrix}$$

Example 2: Now consider subtracting matrix B from matrix A

$$A - B = \begin{bmatrix} 2 & 1 \\ 3 & 4 \end{bmatrix} - \begin{bmatrix} 1 & 0 \\ 2 & 1 \end{bmatrix} = \begin{bmatrix} 1 & 1 \\ 1 & 3 \end{bmatrix}$$

<u>**Scalar Multiplication**</u> (Multiplying a matrix by a number) is shown below:

To multiply a matrix by a single number is very easy:

For example: $3 \times \begin{bmatrix} 1 & 3 \\ 2 & 1 \end{bmatrix} = \begin{bmatrix} 3 & 9 \\ 6 & 3 \end{bmatrix}$

These are calculated as shown:
$$3 \times 1 = 3 \quad 3 \times 3 = 9$$
$$3 \times 2 = 6 \quad 3 \times 1 = 3$$

We call the number ("3" in this case) a **scalar**, and this is called "scalar multiplication".

Multiplying a Matrix by another Matrix

Example 1: Multiply matrix $A = \begin{bmatrix} 1 & 3 \\ 2 & 1 \end{bmatrix}$ with matrix $B = \begin{bmatrix} 3 & 0 \\ 1 & 2 \end{bmatrix}$

Method: $\begin{bmatrix} 1 & 3 \\ 2 & 1 \end{bmatrix} \times \begin{bmatrix} 3 & 0 \\ 1 & 2 \end{bmatrix}$ to work this out you take each number from the first row of the first matrix and multiply it by each number of the first column in the second matrix in order to get the new first element/number of the new matrix shown. You repeat this process for each row and column respectively as shown below. **It is simpler than it sounds!**

$$\begin{bmatrix} 1 & 3 \\ 2 & 1 \end{bmatrix} \times \begin{bmatrix} 3 & 0 \\ 1 & 2 \end{bmatrix} = \begin{bmatrix} 1 \times 3 + 3 \times 1 & 1 \times 0 + 3 \times 2 \\ 2 \times 3 + 1 \times 1 & 2 \times 0 + 1 \times 2 \end{bmatrix} = \begin{bmatrix} 6 & 6 \\ 7 & 2 \end{bmatrix}$$

Multiply each number in the first row of A with each number in the first column of B and add the result as shown by the horizontal and vertical lines. Repeat this process for first row and the second column and so on. **The results are shown above**.

Example 2: Given that $\begin{bmatrix} 2 & a \\ 1 & b \end{bmatrix} \times \begin{bmatrix} 1 & c \\ 2 & -1 \end{bmatrix} = \begin{bmatrix} 8 & -3 \\ 3 & -1 \end{bmatrix}$. Find the values of a, b and c

Method: Multiply the two matrices together in the usual way that is in (i) we multiply each member of the first row in the first matrix by the first column in the second matrix and we get:

(i) $2 + 2a = 8 \implies 2a = 6 \implies a = 3$

(ii)   We carry on the multiplication process (1st row in the first matrix by 2nd column in the second matrix) to get

$2c - a = -3 \implies 2c - 3 = -3 \implies 2c = 6 \implies c = 3$

(iii)   Finally, multiplying 2nd row of the first matrix by the first column of the second matrix we get: $\implies 1 + 2b = 3 \implies 2b = 2$, $b = 1$

So we find that a = 3, b = 1 and c = 3.

Just to check substituting for a, b and c let us multiply the two matrices together: $\begin{vmatrix} 2 & 3 \\ 1 & 1 \end{vmatrix} \times \begin{vmatrix} 1 & 3 \\ 2 & -1 \end{vmatrix} = \begin{vmatrix} 2+6 & 6-3 \\ 1+2 & 3-1 \end{vmatrix} = \begin{vmatrix} 8 & 3 \\ 3 & -1 \end{vmatrix}$ which is in fact the result given in the example above.

Now let us look at Identity and zero matrices.

Identity Matrix: The identity matrix I for a 2×2 matrix is simply $\begin{vmatrix} 1 & 0 \\ 0 & 1 \end{vmatrix}$

The property of the identity matrix is that if you multiply a matrix A with the identity matrix I you get A. In other words A×I = A

Proof: If $A = \begin{vmatrix} a & b \\ c & d \end{vmatrix}$ and $I = \begin{vmatrix} 1 & 0 \\ 0 & 1 \end{vmatrix}$

Then $A \times I = \begin{vmatrix} a & b \\ c & d \end{vmatrix} \times \begin{vmatrix} 1 & 0 \\ 0 & 1 \end{vmatrix} = \begin{vmatrix} a+0 & 0+b \\ c+0 & 0+d \end{vmatrix} = \begin{vmatrix} a & b \\ c & d \end{vmatrix}$

Zero Matrix

This simply when the all the numbers inside a matrix are zero. So for a 2×2 matrix the zero matrix is $\begin{vmatrix} 0 & 0 \\ 0 & 0 \end{vmatrix}$. Yes you have guessed right if you multiply a 2×2 matrix with a 2×2 zero matrix you get a zero matrix

Matrices can also be used for transformations of images

When we want to create a reflection image we multiply the vertex matrix of our figure with what is called a reflection matrix. The most common reflection matrices are:

A reflection in the x-axis is achieved by multiplying the appropriate co-ordinates by the matrix below:

$\begin{pmatrix} 1 & 0 \\ 0 & -1 \end{pmatrix}$

Similarly a reflection in the y-axis is achieved by: $\begin{pmatrix} -1 & 0 \\ 0 & 1 \end{pmatrix}$

For a reflection in the origin we use: $\begin{pmatrix} -1 & 0 \\ 0 & -1 \end{pmatrix}$

(This is the same as rotating the vertices by 180°)

Finally for a reflection in the line y = x, we use $\begin{pmatrix} 0 & 1 \\ 1 & 0 \end{pmatrix}$

<u>We can also rotate and enlarge shapes.</u>

<u>Example 1</u>: Show that P (1, 1) is rotated 90° clockwise about the origin by the matrix $\begin{pmatrix} 1 & 0 \\ 0 & -1 \end{pmatrix}$

Method: multiply $\begin{pmatrix} 1 & 0 \\ 0 & -1 \end{pmatrix}$ by $\begin{pmatrix} 1 \\ 1 \end{pmatrix}$ = $\begin{pmatrix} 1 & 0 \\ 0 & -1 \end{pmatrix} \times \begin{pmatrix} 1 \\ 1 \end{pmatrix}$ = $\begin{pmatrix} 1 \\ -1 \end{pmatrix}$

You can see in the graph below that the **P(1, 1)** has been rotated clockwise by 90° to **Q(1,-1)**

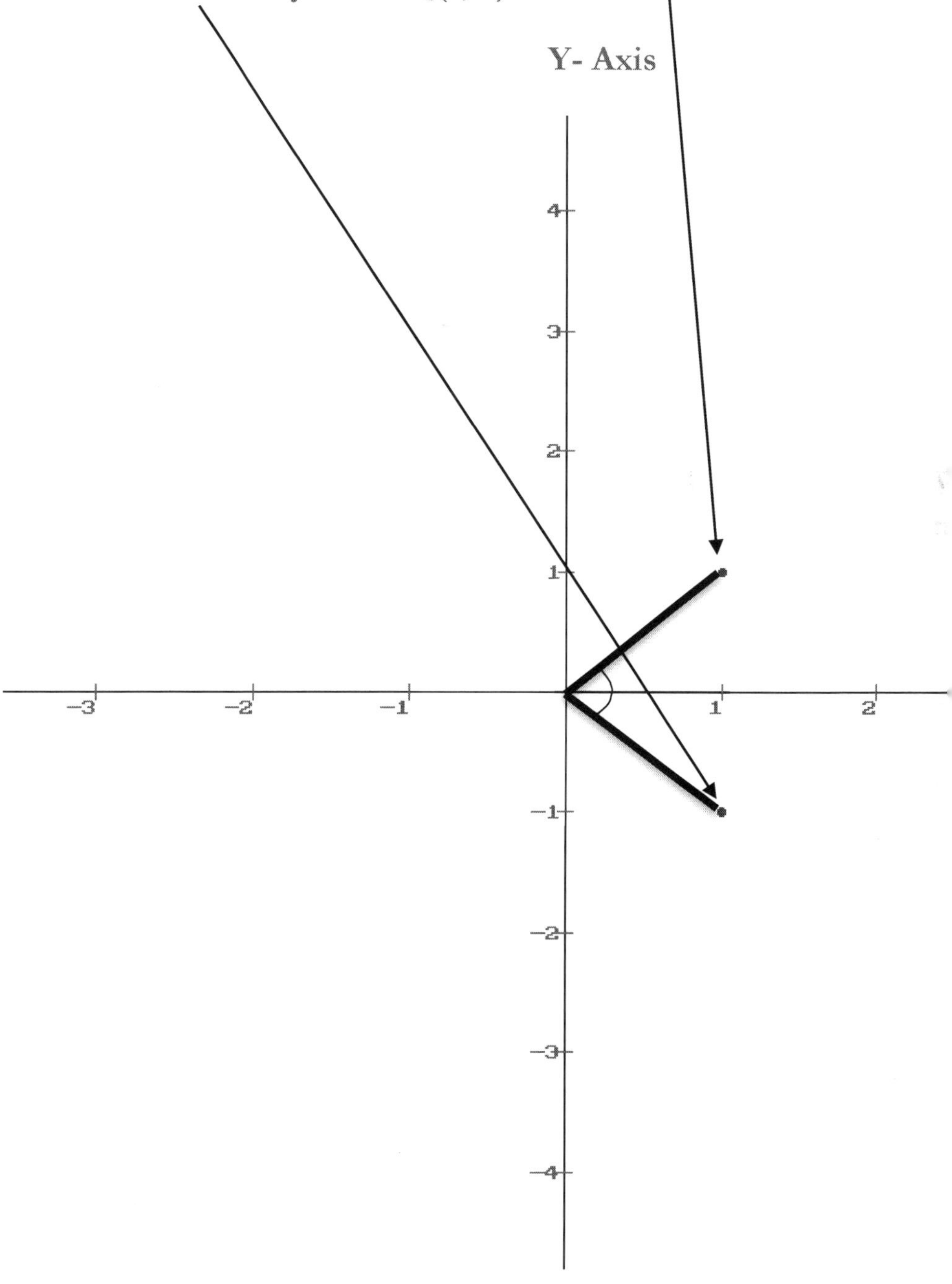

Y- Axis

Example 2: Transforming co-ordinates by two consecutive matrices.

$P = \begin{vmatrix} 2 & 0 \\ 0 & 2 \end{vmatrix}$ and $Q = \begin{vmatrix} -1 & 0 \\ 0 & 1 \end{vmatrix}$

The point A (1, 2) is transformed by the matrix QP to A'. Find the resulting transformation A'.

Method: First work out QP. $\begin{vmatrix} -1 & 0 \\ 0 & 1 \end{vmatrix} \times \begin{vmatrix} 2 & 0 \\ 0 & 2 \end{vmatrix} = \begin{vmatrix} -2 & 0 \\ 0 & 2 \end{vmatrix}$

Now multiply this by A = $\begin{bmatrix} 1 \\ 2 \end{bmatrix}$. So we get A' = $\begin{vmatrix} -2 & 0 \\ 0 & 2 \end{vmatrix} \times \begin{bmatrix} 1 \\ 2 \end{bmatrix} = \begin{bmatrix} -2 \\ 4 \end{bmatrix}$. So the resulting transformation A' = (-2, 4)

Practice Questions on Matrices

(1) If $A = \begin{bmatrix} 2 & 3 \\ -1 & 2 \end{bmatrix}$ and $B = \begin{bmatrix} 4 & -1 \\ 2 & 3 \end{bmatrix}$

Work out:

(a) AB

(b) BA

(c) 3B

(d) A^2

(2) Given that $\begin{bmatrix} 2 & x \\ -1 & 3 \end{bmatrix} \begin{bmatrix} 2 \\ 6 \end{bmatrix} = \begin{bmatrix} 10 \\ 16 \end{bmatrix}$ work out the value of x

(3) If $\begin{bmatrix} 2 & x \\ 3 & 1 \end{bmatrix} \begin{bmatrix} 1 & 2 \\ 3 & y \end{bmatrix} = \begin{bmatrix} 11 & 16 \\ z & 10 \end{bmatrix}$ Find the values of x, y and z

(4) If $M = \begin{bmatrix} 3 & 4 \\ -1 & 2 \end{bmatrix}$ show that MI = M where I is the identity matrix

(5) B(x,y) is transformed to the point B'(-1, 0) by the matrix $\begin{bmatrix} 2 & 4 \\ 1 & 1 \end{bmatrix}$. Work out the values of x and y.

(6) The co-ordinate M (1, 3) is transformed by the matrix PQ to give M'. $P = \begin{bmatrix} 2 & 0 \\ 0 & 2 \end{bmatrix}$ and $Q = \begin{bmatrix} 2 & 4 \\ 1 & 1 \end{bmatrix}$. What is the resulting co-ordinates of M' from this combined transformation PQ?

Answers to Practice Questions on Matrices

1 (a) $\begin{vmatrix} 14 & 7 \\ 0 & 7 \end{vmatrix}$

(b) $\begin{vmatrix} 5 & 10 \\ 1 & 12 \end{vmatrix}$

(c) $\begin{vmatrix} 12 & -3 \\ 6 & 9 \end{vmatrix}$

(d) $\begin{vmatrix} 1 & 12 \\ 0 & 1 \end{vmatrix}$

2   x = 1

3   x = 3, y = 4 and z = 6

4   M×I = $\begin{vmatrix} 3 & 4 \\ -1 & 2 \end{vmatrix} \times \begin{vmatrix} 1 & 0 \\ 0 & 1 \end{vmatrix} = \begin{vmatrix} 3+0 & 0+4 \\ -1+0 & 0+2 \end{vmatrix}$ = $\begin{vmatrix} 3 & 4 \\ -1 & 2 \end{vmatrix}$

5   $x = \frac{1}{2}$ and $y = -\frac{1}{2}$

6   The resulting transformed co-ordinates are x = 28 and y = 8

Printed in Great Britain
by Amazon